indigo

Jenny Balfour-Paul

indigo

BRITISH MUSEUM PRESS

To my husband Glencairn and my children Finella and Hamish

Coming from Elsewhere

What beautiful words arrived here unsolicited
Like refugees. Words such as Indigo,
Whose lingual origins only quaint word-lovers
Trouble to know.

The stuff it stands for, man's first manufactured
Dye, used worldwide from pharaonic shifts
To faded jeans, oldest and loveliest of
The dye-god's gifts.

Beautiful too in its other, natural form –
The sky-god's present, whether neatly folded
in rainbow wrapping or bursting loose
And roughly rolled

In Christmas cloudscapes, relished indeed by lovers
Of colour but viewed mostly with unease,
Arriving unsolicited – like words
Or refugees.

GLENCAIRN BALFOUR-PAUL

First published by British Museum Press
A division of The British Museum Company Ltd
46 Bloomsbury Street, London WC1B 3QQ

First published 1998
First published in paperback 2000

A catalogue record for this book is available from the British Library

ISBN 0 7141 2550 4

Designed by the Bridgewater Book Company
Maps by ML Design
Colour origination by Global Colour Separation, Malaysia
Printed in Slovenia by Korotan

HALF TITLE Japanese indigo gods made by Hiroyuki Shindo. Each year indigo dyers place a new image of the god in their workshop shrine to bring good luck to their dyeing.
FRONTISPIECE Resist-dyed ceremonial cotton shawls, made from strips of European calico embroidered and rope-tied. St Louis, Senegal, acquired in the 1930s.

Contents

Acknowledgements

The subject of indigo is so wide-ranging that I have sought advice and assistance from many quarters over the years, but I would particularly like to express my thanks to all the following:

For fieldtrips outside Europe, I owe much to many people for their hospitality and help. In Thailand, China and Japan: Ramsay and Stella Melhuish, Rosalind and William Haddon, Gina Corrigan, Zaou Lu, Satoshi Ushida, Osamu Nii, Hiroyuki and Chicako Shindo (and Hiroyuki's indigo gods), Betsy Sterling Benjamin and the dyers of Tokushima. In West Africa: Hilton and Anita Whittle, Adama Mara, Sarah Brett-Smith, Aminata Traouré, Abdoul Aïdara, the late Nancy Stanfield, Violet Diallo, Caroline Hart, Bernard Gardi and Nike Olaniyi-Davies. In India: Padmini Tolat Balaram, staff at the National Institute of Design in Ahmedabad, Lotika Varadarajan, Vickram Joshi, David and Jenny Housego, Judy Frater and the Khatri family at Dhamadka. In the Near East and North Africa: all the friends, colleagues and those involved in the indigo industry mentioned in my book *Indigo in the Arab World*. It is impossible to name all the indigo growers, dyers and related textile craftsmen I have visited in all these countries, but I am grateful to them all for sharing their expertise.

For information on Central and South America I am indebted to Chloë Sayer, Martha Turok, Jamie Marshall, Ann Hecht, Ana Roquero and Cati Ramsay; on Indonesia, to Sandra Neissen, Jan Wisseman Christie, John Gillow and Traude Gavin; and on the Solomon Islands, to Virginia Bond.

As for Europe itself, my special thanks for general help and encouragement go to the late Susan Bosence, Dominique Cardon and Jacqueline Herald, and for assistance in various directions to Ruth Barnes, Joss Graham, Loan Oei and Graham Ashton. I have been helped in Austria by Regina Hofmann and Josef and Elisabeth Koó; in Germany by Hansjürgen Müllerott and Michael Bischoff; and in the UK by Norman Wills, Jill Goodwin, Joan Thirsk, Su Grierson, Gigi and Roddy Jones, Levi Strauss Co., Gerald Harrop, Robert Fox, Ros Hibbert, Mike Quinnin and Iris Cleatheroe. Those who have put me straight on matters chemical include Penelope Rogers, Jan Wouters, Jo Kirby, Chris Cooksey, Ken Seddon and Jurek Mencel; and on botanical matters Brian Schrire, Martyn Rix, James Morley, David Hill, Kerry Stoker and David Cooke. At the British Museum, Shelagh Weir, Imogen Laing, Jim Hamill and Hans Rashbrook stand out for their helpfulness; and at the British Library's Oriental and India Office Collections, David Jacobs, Anthony Farrington and Beth McKillop.

At Exeter University I have received unfailing support from Kamil Mahdi, Jennifer Davis, Ruth Butler, Bobby Coles, Parvine Foroughi, Lindy Ayubi, Sheila Westcott, Heather Eva and Paul Auchterlonie.

For reading and commenting on various sections of my manuscript I should especially like to thank experts Chris Cooksey, Jo Kirby, Eddie Sinclair, Brian Shrire, Martyn Rix, and above all Jacqueline Herald for tackling the whole draft. However, any errors and omissions are entirely due to my own shortcomings.

It has been a great pleasure working with the friendly staff at British Museum Press – especially Rochelle Levy, Alasdair Macleod and Susanna Friedman, but above all Coralie Hood, unflappable editor of this book. My debt to Colin Grant for his remarkable patience over the text is also enormous.

I am most grateful to Exeter University for awarding me an honorary research fellowship since 1993, and two research grants. Other generous research grants from the following organizations financed my overseas fieldwork, and attendance at specialist conferences, between 1983 and 1997: British Academy, Royal Society, Nuffield Foundation, Pasold Research Institute, INTACH (Indian National Trust for Art and Cultural Heritage), Universities' China Committee in London, Elmgrant Trust, David Canter Memorial Fund, Gilchrist Educational Trust, the Dyers' Company, London, and the Marc Fitch Fund.

Finally, I wish to acknowlege my long-suffering family. Finella and Hamish spent part of their childhood being dragged from one indigo workshop to another, as far apart as Thailand and Morocco. My mother, Jill Scott, has been supportive in many ways, not least by tolerating my woad and indigo plants in her vegetable garden (as too did Bruce Dimmick). But above all the constant support in every way of my husband, Glencairn, enabled the entire project to be realized.

Preface

Young woman wearing a jacket of beaten indigo-dyed cotton, carrying a basketful of cloth just dyed in indigo to rinse in the river. Judong, south-west China, 1993.

One summer's day in 1981, in the courtyard of a Devon farmhouse, I first plunged my hands into an indigo dye vat, pulled out a length of cloth and watched the apparently magical transformation of yellow to indigo blue taking place before my eyes. I was hooked, and resolved from then on to discover more about this elusive and most important of all dyestuffs. (When, like Alice, I dived down the rabbit hole, I had no idea what curiosities would be revealed.) This resolution, reinforced by several years' personal experience of the Arab world, whose age-old indigo industry was about to disappear largely unrecorded, led to my initial study on the Red Sea coastal plain of Yemen. This research expanded to encompass the rest of the Near and Middle East, and the outcome was a doctorate at Exeter University and subsequent book. As I had already been making comparisons with the indigo industry in other parts of the world, I needed little encouragement from the British Museum Press to continue this study. Apart from the obvious library and museum research, it led to further field trips to China, Japan, Thailand, India, West Africa and many parts of Europe, as well as to my own experimentation with indigo plants and dyes. This book, an attempt to present all the various absorbing facets of indigo in a worldwide context, is the result.

I have seen my role rather as that of a spider in the centre of a web, feeding from all sides. The subject embraces so many disciplines in the sciences and the arts – botany, chemistry, ethnography, economics, medicine, art, textile and social history amongst others – that no one could be an expert on them all. I have drawn on information provided by many specialists with different angles on the subject to supplement my own first-hand experiences. Dye conferences in particular have provided not only an invaluable cross-fertilization of ideas, but a more down-to-earth exchange of seeds, plants and other specimens and samples. Innumerable textile researchers and anthropologists have uncovered the complex world of cultural values hidden within dyed and patterned cloths. After more than fifteen years of study I remain captivated by a subject that appears to be inexhaustible, for there are always new threads to unravel, and more paradoxes to try and penetrate. In particular the ways people from very different cultures have empirically resolved so many of the intricacies connected with indigo production are a source of wonder. It has been painful, within the confines of a single book, to omit a good deal of historical and technological detail, but I hope the notes and bibliography will compensate and provide a stimulus for those wishing to learn more.

My overriding aim is that the high-tech, turn-of-the-century reader will share my fascination with a natural product which provided our ancestors with a world of blue, and will marvel at a not-so-distant past when every colour used by the dyer was supplied by nature without recourse to the chemist's laboratory. I hope too that the production of indigo from natural sources, although inconceivable today on a large scale, will be kept alive by those who appreciate its special qualities and its intriguing story.

Of blues there is only one real dye, indigo
(WILLIAM MORRIS, 1834–96)

Introduction
The Myth and the Magic

Tracing the history of almost any staple commodity as it has threaded its way through history is bound to be a revelation. Whether the commodity is as basic as sugar, as dubious as opium or as mundane as the humble potato, when investigated closely it will be found to have had widespread ramifications and to have touched on many aspects of life in divers cultures. Indigo is one such commodity, cropping up in many fields of science and the arts: agriculture, economics, botany and chemistry, trade and industry, ethnography, costume, furnishings, the applied arts and even medicine and cosmetics. But in addition to the intriguing history of the subject, the aesthetic and technical aspects are often quite extra-ordinary. Its unique production methods with their spiritual associations, and the beauty of the colours produced on a wide range of textiles, have lent an aura of mystique which still lingers on today. And the universal adoption of indigo-blue jeans has united cultures worldwide.

Until recently, most textile and carpet experts have focused their attention on weaving, embroidery and design. Even though colour is usually the first thing to strike the eye, and is the most definable and dominating aspect of a textile's beauty, the means of producing colour and creating coloured fibres have often been taken for granted. The skills of the weaver or embroiderer are indeed a source of wonder, but so too are the transformations wrought by the dyer. And an understanding of the

Lengths of indigo-dyed cotton hanging in the courtyard at Bayt Muhammad Ali Abud, Zabid, Yemen in 1983.

manner in which colours have been obtained, processed and applied to all types of fibres throughout history increases the appreciation of textiles as well as being essential for the conservation of historic pieces. The choice of dyes and the status of dyers can help to illuminate the story of many cultures (particularly those with no written documentation). The colours used on textiles and artefacts, their social significance and the scope of their trade, are part and parcel of a people's overall history.

It is only during the twentieth century that synthetic dyestuffs, invented in the second half of the previous century, became widely available. Before that, for well over four millennia, all dyestuffs were made from natural ingredients found mostly in the plant kingdom, with the exception of the important red insect dyes (kermes, cochineal and lac), some metallic oxides and the renowned shellfish purple dye. It is a strange fact that green, the colour of plants, is the one dye colour not obtainable from them. Innumerable plants yield yellow and brown dyes. These, along with the reds and blacks, belong to various groupings classifiable by their chemical structure. Indigo and its close relation shellfish purple are chemically in a class apart. They form the extraordinary 'indigoid' group, whose production methods are so intriguing that they still tantalize today's organic chemists. The word indigo refers to the blue colouring matter extracted from the leaves of various plants including woad. Thanks to its unusual chemical make-up, indigo can be treated both as a dyestuff, in which case cloth or yarn is immersed in the dye vat, or as a blue pigment (sometimes referred to as 'indigotin') for paints and inks. European writers have tended to distinguish between imported indigo dyestuff made from tropical or subtropical indigo plants and that produced from home-grown woad plants, referring to the latter quite simply as 'woad'. Elijah Bemiss, for example, observed in 1815: 'We are not furnished but two drugs that give a permanent blue, and they are, indigo and woad.'[1] The reason for distinguishing the two is that traditional processing of woad leaves produced a low indigo yield suitable for dyeing absorbent wool fibres, whereas the foreign indigo plants yielded much higher concentrations which were ideal for dyeing the less absorbent vegetable fibres such as cotton and flax. But chemically speaking all produce the same indigo blue. No other dyestuff has been valued by mankind so widely and for so long. One of the world's oldest dyes, it remains the last natural dye used in places that have embraced synthetic dyes for every other colour.

One of the eternal mysteries relating to dyeing with indigo is how did it all begin? This question can never be fully answered. Until recently a view prevailed that many textile technologies, including that of indigo dyeing,

Franco-Flemish tapestry, *c.*1500, of *The Hunt of the Unicorn*. Here the unicorn dips his horn into the stream to rid it of poison. Silk, wool and silver-gilt threads (3.68 x 3.78m).

dispersed outwards from India, the 'home' of textiles. The influence of Indian techniques in general can hardly be overestimated, but it is now accepted that in the distant past people discovered how to dye fibres blue with the various indigo plants independently at different times and in different places, even in those which later embraced from elsewhere new species and more efficient production methods to meet an expanding demand. Given that dyeing with indigo is a complex procedure, and that different plant sources require different treatments before they will yield their dye, it is remarkable how many civilizations appear to have discovered the secrets on their own. Perhaps some damaged indigo leaves were soaked with stale urine or ash liquid left over from a wood fire; both could have coloured fibres blue under favourable conditions. The puzzling origins of indigo are embedded in numerous ancient legends, suggesting a role in indigenous textile cultures stretching back over several thousand years. Southeast Asian mythology, for example, is full of indigo stories.[2] In Liberia a tale is told of the way post-menopausal women gained from the High God the secret of blue dyeing with indigo thanks to a seeress who broke off a piece of blue sky to eat (after which the sky was pulled up high out of reach).[3]

Even allowing for changing visual perceptions over the centuries, all the discoverers of indigo dyestuff must surely have been amazed and delighted by its colour, for most available hues in the ancient world are related to the earth palette. Cave painters, for example, were limited to mineral pigments of red, ochre, brown, black and white. The rareness of blue dyes compared with yellows is echoed in the vegetable kingdom at large. Blue flowers are much less common than yellow for example, and edible blue substances are almost non-existent. So finding that nature could produce a colourfast blue dyestuff by a process akin to alchemy must have been an extraordinary revelation, whose impact is hard to imagine in today's multi-coloured world. Once discovered there was certainly no looking back.

Now that natural indigo plants are only grown on a small scale, whether for the traditional practitioner, the hobby dyer or for modest commercial revival, it is hard to envisage the extent of the indigo industry in the past, or the effects it has had on all kinds of social, cultural, political and more concrete aspects of community life. This study will, among other things, span both the socio-economic aspects of continuing domestic usages at village level and the global scale of former commercial enterprise. As a commodity traded between the continents by land and sea, indigo produced a substantial income for the countries dealing in it long before it featured as a hugely profitable field for colonial exploitation. Many aspects of life, including lasting monuments such as the fine woad

Two women in everyday indigo-dyed clothing. Near Zhengfeng, south-west China, 1993.

merchants' houses which dominate the elegant centre of French Toulouse, many grand buildings in Calcutta, and a school and library in America's Georgetown have profited from woad or indigo wealth. Income from woad grown on their estates even contributed to the fortune of the Spencer ancestors of the late Diana, Princess of Wales.[4]

The catch phrase 'universal' has frequently been applied to indigo, and with just cause. Not only do indigo-bearing plants grow in most corners of the globe, in both tropical, subtropical and temperate climates, but the dyestuff extracted from them has a unique chemistry which renders it compatible with every type of natural fibre, both animal and vegetable, unlike other natural dyes. Indigo is one of the most colourfast of natural dyes, well demonstrated by the countless medieval and Renaissance tapestries exposed to daylight over the centuries, whose former green foliage has reverted to blue due to a process known as 'preferential fading' or, unflatteringly, as 'blue disease'. Yellows are by nature notoriously fugitive and most other colours tend to become duller as they fade. Indigo blue always retains its beautiful hue even if it grows paler, hence the only original colour of the Bayeux tapestry that remains true is the indigo blue of its woad-dyed wools. Perhaps most importantly of all, the range of colours relying on indigo was enormous, as it was required for much of the spectrum as well as blue. Purple dyes are scarce in nature, and most blacks were unsatisfactory, based as they were on corrosive metallic compounds; but indigo, in combination with yellow, red and brown dyes, could provide these shades and greens.[5] The medieval European dyer used woad in the creation of most colours, and even in the early eighteenth century a German text declares of woad that 'the Colour … is always good and the best and most necessary Ingredient for Dying [*sic*], since 'tis used in the Composition of most Dyes, which can neither be rendered good or lasting without it'.[6] Late nineteenth century European dye manuals still describe the indigo/woad vat as essential for 'bottoming' cloth,[7] and dyers worldwide have considered that a final dyeing with indigo would generally make fabrics more colourfast and longer lasting. One Indian dye book describes the production of endless subtle colours ranging from 'regal purple' to the 'colour of kidney beans' and 'light canary', all of them made in Gujarat from indigo combinations.[8] There is no other dyestuff with such a comprehensive application.

The double attraction of indigo as a clothing dye has always been a combination of practicality and widespread availability. We talk of patches of blue sky being the size of 'Dutchman's trousers' and refer to 'blue-collar workers' because the standard dress for the working people of Europe, as of China, Japan and many other countries, was tailored from cloth dyed in indigo. In many societies the native word for indigo still refers primarily to the substance, rather than to its colour. This was also the case in the West before Sir Isaac Newton, in 1672, fused his theories on musical and colour harmony by introducing orange and indigo to the five more prominent colours of the spectrum (a decision that has aroused much controversy).[9] However, over the centuries indigo has accumulated layers of significance and symbolism, although there may be a tendency in the West to exaggerate or even misinterpret these aspects. Some symbolic associations do indeed relate to its actual colour, but many others arose out of its chemical mysteries and durability. Any country woman could gather plant material from the surrounding countryside, boil it up with some wool and a pinch of alum and produce warm earthy shades with varying degrees of fastness. But the only way she could obtain colourfast blue yarn or cloth was by resorting to the specialist, who guarded his or her secrets well. Even where people did learn to use indigo plants at a domestic level, they had to acquire a particular expertise to ensure satisfactory results, for indigo dye behaves in a remarkable, sometimes seemingly capricious, way. This is a main reason why indigo became a special, and often expensive, dye around which many myths, superstitions and religious rituals evolved over the centuries. These will be covered in subsequent chapters, as will medicinal, cosmetic, artistic and other usages (from Classical times to the present), which frequently intertwined with indigo's other qualities to enhance its esteem and mystique.

Bedouin tribal rug of Jordan (detail).

In many cultures indigo blue has of course been equally valued for aesthetic reasons, and continues to be so, although here too its symbolism can be as mystifying as the dye bath itself. It echoes the infinite richness of the sea, the midnight sky, the shadowy dusk and early dawn, and represents the elusive seventh colour of the rainbow which some people simply cannot see. In the medieval and Byzantine worlds blue was associated with divinity and humility, and in India with infinity and the capricious god Krishna. Many see it as a spiritual or reassuring colour, standing for loyalty, as opposed to yellow, the colour of cowards. Although indigo represented happiness for the craftsman William Morris, for many it evokes 'the blues', both in mood and in music. 'Mood Indigo', Duke Ellington's immortal song, provokes a sense of blue melancholy also expressed by Goethe, Picasso and others. Some societies, including Islamic and Indonesian, which consider indigo 'dark', or 'black', have linked indigo directly with 'black magic'.

In addition to serving as trade goods and gifts, textiles have always been used as a channel for cultural and religious expression, particularly in ethnic communities with no written language. Consequently the producers of special textiles, often women, have been able to gain

Sharada Utsava *pecchava* (painted hanging) of Krishna playing the flute to enchant the *gopis* (milkmaids). Pale blue and gold on dark indigo-dyed ground. Deccan, late eighteenth century (detail).

economic, social or even political power thanks to the expertise required to make them.[10] Indigo, with certain other dyes, has played a major part in the production of textiles imbued with symbolic power. Garments which included dark indigo-dyed threads, indicating wealth and therefore, by extension, prestige, were used for important burials in, for instance, Ancient Egypt, Palestine, Mali and Peru. No one knows when indigo-dyed cloth first became involved in mourning rituals in cultures as far apart as those of West Africa and Indonesia, but clearly the transformations intrinsic to the dye vat were perceived as reflecting the spiritually transforming rites of passage of life itself. For this reason, in parts of Indonesia and elsewhere where synthetic dyes are now the norm, natural dyes are still a requirement for ceremonial textiles. Tribal chieftains of Africa, the Middle East and South America all wore dark blue indigo-dyed robes with pride, as did members of the ruling classes of China, Japan and Indonesia. Specific indigo-dyed textiles, such as the turban, have long been used as a mark of initiation, reward or status. On the other hand some, such as faded blue work clothes, have been chosen deliberately as an anti-establishment political statement: most notably, in their earlier days, blue denim jeans, the most widespread and enduring post-war sartorial statement.

Whatever its associations, indigo blue is cool and relaxing, counter-balancing the warm end of the spectrum. It contrasts with certain other colours to impressive effect; in textiles the red/blue and blue/white combinations in particular have always been popular. What could be more harmonious than the blend of madder reds with indigo blues on many central Asian tribal rugs or the morinda reds and indigo blues of an Indonesian ikat? The fashion for blue and white, expressed famously in porcelain, could also be indulged, thanks to indigo, in textiles. There could be few more striking sights than a tall Nigerian woman enveloped in a boldly patterned blue-and-white *adire* cloth or a Japanese woman in an ikat kimono. Not only is its colour satisfying, but the singular chemistry of indigo's dyeing processes, whether using natural or synthetic indigo, allies exceptionally well with many popular textile-patterning techniques,

Cotton warp ikat with birds and stylized tree designs in morinda reds and indigo blues. Sumba island, Indonesia, late nineteenth century.

Woman and her baby wearing an indigo stitch-resist wrap of an 'old African' design, standing beside a field of guinea corn. Nigeria, 1963.

Weavers dressed in *kasuri* (ikat) clothing, with *kasuri* tied yarn in the background. Orihime, Japan, *c.*1965.

notably those based on a 'resist', or 'reserve', process. And even when it ages indigo retains, again thanks to its special chemistry, its inimitable blue. The Japanese cultural appreciation of the aesthetics of age places great value on the special qualities of faded indigo[11] – as does anyone who wears blue jeans. And it is the Japanese, whose culture manages to combine modern high technology with traditional material culture, who elevate indigo growers and dyers to the status of 'national living treasure'.

Even with the huge range of synthetic dyestuffs available to the industry today, the fashion pages continue to extol the unique and timeless attractions of indigo-dyed clothing. To fulfil this constant demand the manufacturers of synthetic indigo, produced in a different way from all other modern dyes, play a major part in the continuing story of indigo. As the chemical company Zeneca's promotional literature puts it: 'Indigo continues to survive and flourish in the face of severe competition from today's most sophisticated dyestuffs as it continues to play a unique role in serving a textile market that is both ancient and modern, linking technologies that are centuries apart.'

For the researcher there is an abundance of information on certain aspects of indigo production, particularly concerning its commercial exploitation in the colonial period, but a paucity of sources, or even downright misinformation, on others. For example, botanical identification often needs revising, apart from the ongoing re-classification of plant species which applies directly to some of the indigo-bearing genera. Dye processes have sometimes been recorded by those unfamiliar with the unusual chemistry of the indigo vat so that, to give a specific example, mordants are often assumed to be required, as they are for most dyes, even though a defining feature of indigo dye is that no mordant is needed. Furthermore, indigo dyers the world over have been reluctant to yield up their hard-earned secrets to outsiders, for fear of rivals or of the wrath of malevolent spirits. Historical revisionism can be disquieting too. Julius Caesar's lines about the Ancient Britons painting themselves with woad to appear fierce in battle are much quoted, but the translation of vitrum as 'woad' is questioned by some. Marco Polo is a much-cited source for the first detailed description of the production of indigo in north-west India, yet one scholar has recently suggested that the famous Venetian traveller actually composed his so-called 'travels' from a compendium of existing written sources.[12] If his descriptions of indigo production were indeed relayed at first remove, they must lose some of their edge.

As the world's interest in natural dyestuffs increases, and historical examples are more rigorously scrutinized, more information emerges to revise and extend our knowledge. The story of indigo, in its various guises, will surely never end.

[Send] a man cunning to work
in gold . . . and in purple, and crimson, and blue
(KING SOLOMON TO THE KING OF TYRE)

From Antiquity to the Middle Ages

Indigo has had an exceptionally long innings in the fields of industry and commerce. The last two millennia of indigo's economic history are neatly encapsulated in its names. The word indigo itself derives from the Greek *indikon*, Latinized *indicum*, originally meaning a substance from India, indicating the import of indigo pigment by the Graeco-Roman world. The Sanskrit word *nila*, meaning dark blue, spread from India eastwards into Southeast Asia and westwards to the Near and Middle East, probably both through pre-Islamic trading routes and with the subsequent trade diffusion of the product in the Islamic era.[1] The Arabs conveyed their word *nil*, or, with the definite article, *an-nil*, further west in the course of their conquests across northern Africa and into southern Spain. Subsequently the Spanish and Portuguese transmitted the words *añil* and *anilera* to Central and South America in the sixteenth century. In that century, in the reign of Queen Elizabeth I, a British Act of Parliament referred to indigo as 'nele, alias blew Inde', while European travellers' and merchants' reports interchange *neel/anyle* with *indico*. In the seventeenth century indigo became the common name, but a reminder of indigo's etymological history survives into modern times with the word for a class of dye, 'aniline', so named because indigo formed the basis of the first dyestuff to be chemically synthesized (see the Glossary).

Stained glass window of woad dyers in the medieval Barfüsserkirche, Erfurt, Germany.

This chapter will look at the evidence of the earliest uses of indigo dye produced from tropical and subtropical indigo plants as well as from temperate woad plants, and will consider its widespread importance during Europe's Middle Ages and equivalent periods elsewhere before the wholesale changes brought about by European colonial expansion. Specific uses in a wide variety of textiles throughout history will be covered in Chapter Seven.

Inevitably the further back in time that dyestuffs are traced, the more reliance has to be placed on archaeological sources, and unfortunately organic substances (such as vegetable dyes and the materials they coloured) are the most perishable. Nevertheless, more systematic examination of textile finds and of botanical specimens is gradually fitting together more parts of the jigsaw.[2] Although the dating of fragments is a different issue from identifying dyestuffs found in them, the two aspects can sometimes throw light on each other. Accurate dating remains a particular minefield, often requiring revision as new evidence emerges. There is always the danger that archaeological excavations will create false emphases, based as they often are on areas with elaborate burial customs and where arid or anaerobic climatic conditions have favoured the preservation of artefacts. Furthermore, the survival of dyed fibres depends not only on surrounding conditions but also on the ageing effects of the various dye ingredients themselves. This arbitrary selectivity was until recently often reinforced by a tendency among archaeologists to undervalue, or even discard, textile finds, and among textile historians to take colours for granted without exploring their source. Archaeological evidence is sometimes underpinned by early written sources or visual material such as wall paintings, but certain aspects are bound to be speculative and there is still much to discover.

Excavated fragments reveal that indigo dyestuff, from whatever plant source, was already in use by the third millennium BC. In general the identity of dyestuffs used on ancient artefacts cannot be deduced by colour, as this alters owing to changes over time, and in any case dyes were often combined to create particular colours. Where indigo was used alone, however, it does retain its blue hue, even if it fades, and its presence can be simply established by scientific analysis. But despite increasingly sophisticated analytical equipment and methods, dye chemists are unable to identify the plant species used on dyed fibres (or even to distinguish natural from synthetic indigo, although carbon-14 techniques are getting close). In future more may be learnt by examining other compounds, including contaminants, present in a sample.[3] The botanical source can, however, sometimes be guessed at by the provenance and date of a sample or by the analysis of other natural dyestuffs present whose dating and provenance are known.

The ancient world

Decorative bodily adornment has its roots in earliest history. At some stage different prehistoric peoples invented the weaving loom, and they coloured woven cloth and other materials by applying to their surfaces pigments (with a binding medium) and other substances which were not generally colourfast. The harder skill was also acquired of actually *dyeing* threads, which involves creating a chemical bond between fibre and colouring substance in solution. Dyed threads could then be inserted into weavings for more permanent decorative effect than could be achieved with surface pigments. Considering the complexities of dyeing with the indigo-bearing plants it is remarkable to find its early use by very diverse and geographically separate civilizations.

Textile fragment from Deir el-Bahri, Upper Egypt, with indigo-dyed stripes. Eleventh dynasty (*c.*2000 BC).

Egypt

In Ancient Egypt, where blue colours were revered, many archaeological textile fragments have been preserved thanks to the dry climate, sterile sand and local burial customs. The country's woven textiles were widely esteemed, and weavers began to insert rare blue stripes into the borders of plain linen mummy cloths from the Fifth Dynasty (*c.*2400 BC), probably because only indigo dye is well absorbed by flax fibres. Indigo-dyed linen and occasionally wool (a much easier fibre to dye) are more commonly found, sometimes as part of multi-coloured patterns, in textiles dating from the Middle Kingdom (from *c.*2040 BC) and particularly the New Kingdom (from *c.*1560 BC). The celebrated funerary wardrobe of Tutankhamun, for example, includes a state robe that is predominantly blue, and other garments and embroidery threads with some blue.[4] Wall paintings, most famously that in the tomb of governor Khemhotep at Beni Hasan, *c.*1900 BC, show a taste for bright decorative clothing (see p.16).

There is much speculation about the source of the indigo dyestuff used for Egyptian and other early eastern Mediterranean textiles.[5] Indigo dye made from indigofera species could conceivably have been amongst the luxury goods traded northwards up the Red Sea from Punt or from southern Arabia, possibly even by the third millennium BC.[6] Southern Arabia, with its incense revenues and pivotal trading position straddling the Indian Ocean and the Red Sea and East Africa, was a wealthy region whose renowned spice and textile trade could have included indigo, either locally produced or transported from further East. But even if this was the case, its import costs would probably have excluded its use in Egyptian dye vats. It is botanically possible that local indigofera species could have been used for blue dye by the Ancient Egyptians, but it is most likely that they made use of the woad plant, *Isatis tinctoria* L., indigenous in parts of north Africa as well as in Europe and western Asia, even though it is much less efficient as a source of indigo dye than the tropical and subtropical indigo plants, particularly for dyeing cellulosic fibres. Hieroglyphic inscriptions, such as those at the largely Ptolemaic temple of Dendera,[7] which show the Egyptians' appreciation of several different shades of blue, mention plants producing a blue colour resembling lapis lazuli, but botanical identification remains hazy.[8]

Whichever dye source was used, the Egyptian dyer was clearly highly skilled long before the appearance of written evidence from Hellenistic Egypt in the form of second-to-third-century AD Greek papyri, particularly the *Holmiensis*, or 'Stockholm', papyrus which describes harvesting and dyeing with woad.[9] A urine fermentation method was employed (see p.124), adding local soapwort (still being used by Omani indigo dyers in the 1980s), and often overdyeing with madder or lichen dye to imitate the shellfish purple so highly sought after by this time.(Even today people of Oaxaca in Mexico will taste purple-dyed fabric to ensure they are not being palmed off with an indigo and red fake.)[10]

Shellfish purple

It would be absurd to discuss indigo in the ancient world, particularly the eastern Mediterranean, without mentioning the closely related shellfish dye which was extracted from species belonging to the *Muricidae* and *Thaididae* families. This was famously used by aristocratic Phoenicians, Romans and Byzantines, hence the common epithet 'Royal', 'Imperial' or 'Tyrian' (for Tyre produced the best quality). Purple dye became the trademark of the Phoenicians. It reached its apogee as a status symbol during the Roman and Byzantine empires before its relegation to the sidelines of dye history following the Turkish conquest of Constantinople in 1453. Thereafter knowledge of its production methods vanished. No one knows whether the technology for indigo dyeing preceded that for shellfish purple or *vice versa*, as both seem to date back to the third millennium BC.[11]

Despite shellfish being of animal origin, as opposed to the vegetable indigo, the chemical composition of the extracted dyestuffs differs by just one or two bromine atoms[12] and until fairly recently they were sometimes confused, as indigo is the basic component of both. For instance, ancient indigo-dyed textiles which had been rinsed in sea water or buried in saline soil (both of which contain bromine) have sometimes been mistakenly analyzed as dyed with shellfish purple. Moreover plant indigo itself can appear purplish due to the presence of natural red indirubin (see p.132). References to dyed textiles in the Old Testament are ambiguous, as they do not describe precise colours or dye substances. However, it is generally accepted that the Hebrew word *argaman* refers to a reddish-purple colour while *tekhelet* was used for blue or blue-purple, although standard translations of the Hebrew are imprecise. The exact shades may never be known since, despite all the literary references, extant textile samples are rare. Shellfish purple residue has, however, been found on potsherds of Ancient Israel thought to have formed clay dye vessels.[13] Dyed textiles had great significance in Jewish religious ritual. Hebrew law, for example, prescribed the wearing of *tekhelet* fringes and tassels, which are presumed to have been dyed with shellfish purple. The main attraction in biblical times was probably not so much colour as prestige relating to its animal origins and its high cost. Even today, for those of orthodox Jewish faith, investigation into this mysterious dyestuff often has a religious rather than an aesthetic motivation.

Pliny gives an incomplete account of production in Roman times[14] and Vitruvius notes the effects of sea temperature on the final colour.[15] Dyeing methods, climatic conditions and seasons would certainly have affected the outcome, but it is generally accepted that in the Mediterranean the main reddish species were *Murex brandaris* and *Purpura haemastoma* while the hermaphrodite *Murex trunculus* produced the bluer hues. Indeed, as well as purple this mollusc apparently contains pure indigo.[16] Some scientists have suggested that this may be due to a sex change, the bluer colours being produced during the masculine phase. (Dare one suggest this as a possible origin of the expression 'blue for a boy, pink for a girl'?)

Huge quantities of shellfish had to be killed (about 10,000 to obtain one gram of dye) in order to extract from the tiny hypobranchial glands enough of the photo-sensitive whitish secretion containing the precursor to purple. Beaches in Tyre and Sidon as well as Crete are still piled with heaps of left-over shell deposits. (In Central America, where the ancient practice of shellfish purple dyeing still survives today, the glands of *Purpura patula* are 'milked' without harming the whelks, and used directly on yarn wetted in sea water.)[17] Shellfish purple dyers, like those for indigo, needed specialist knowledge. Many dye chemists today are experimenting with purple shell-fish,[18] but no one has yet established the precise process by which the ancients used this elusive dye. Moreover, emulating the scale of their dyeing operations is inconceivable today.

Ancient Israel and Palestine

Archaeological textile fragments, and warnings in the Talmud (the fundamental body of Jewish laws and traditions codified around AD 200) against counterfeit shellfish purple, reveal that a fake purple colour was created by combining indigo with a red dye, usually madder. These two, used both separately and combined, are in any case the main dyes found on textiles of Ancient Israel and Palestine, and indigo was also combined with other dyes to produce greens and blacks. Indigo makes its appearance here from the end of the second millennium BC; as in Egypt, it is the only colour found on the earliest dyed linen fragments. Examples from Kuntillet Ajrud, thought to be Iron Age, already feature indigo, but it becomes much more common on later finds, of wool as well as of linen. This is especially the case from around the beginning of the Classical/early Christian period, in such finds as those from the desert sites of Masada, Qumran and the Cave of Letters (where madder and insect red, saffron and tannins have also been identified).[19] Symbolic geometric patterns in blue decorate scraps of the linen wrappers used to encase the urns which contained the Dead Sea scrolls.

Semite women of the Aamu tribe of southern Palestine. Tempera copy of a detail from a wall painting in the Beni Hasan tomb of Khnemhotep, c.AD 1900.

Part of a cuneiform tablet, *c.*600 BC, from Ancient Babylonia, which gives the world's earliest written instructions for wool dyeing, including indigo recipes.

Western Asia

There are other hints and pieces of evidence to suggest a widespread early use of indigo. In theory blue dye (from woad) could have been used as early as around 6000 BC in the Near East, for evidence at the early Neolithic site of Çatalhöyük in southern Anatolia suggests knowledge of dyes at that time.[20] Coloured textiles of the third millennium BC have been found in Anatolia. Babylonian texts of the early second millennium from the palace halls of Mari on the Euphrates mention 'garments dyed in blue', and the textual evidence is supported by numerous fragments of wall paintings depicting several shades of blue.[21]

At a later date in the region there is ample evidence of the use of indigo, including traces of indigo pigment decorating a Roman parade shield discovered at Dura Europos, just upstream from Mari. Indigo resist-dyed cotton and the more usual woollen dyed yarns for the pile textiles have also been found among the thousands of specimens, dated late second century BC to fifth century AD, excavated in the at-Tar burial caves of south-western Iraq, a Silk Road trading centre.[22] Analyses of first-to-third-century AD textile fragments found in tombs at Palmyra further north-west, reveal shellfish purple, an indigo/kermes (red) combination and plain indigo.[23]

The most exciting recent development concerning early dye history is the discovery in 1993 in the British Museum of another section of a neo-Babylonian cuneiform tablet, dated to the seventh century BC. Joining the two clay fragments together enabled about half of a seventy line text to be deciphered. This unique tablet contains the first known recipes for dyeing wool.[24] Of particular interest are the recipes for both base colours and colour mixes. One clearly describes indigo dyeing, with its characteristic repeated dippings and airings necessary to produce 'lapis-coloured wool' (*uqnatu*), but does not mention the dye source (Indian indigo may have reached Mesopotamia by this time).[25] Other recipes describe overdyeing to produce reddish and bluish purples (using the words *argamannu* and *takiltu* – the similarity with the Hebrew terms is striking).

Yemen has a long history of dyeing with indigo and also as a centre for the East-West luxury goods trade.[26] These may have included dyed textiles long before the Christian era. There are tantalizing references to 'blue cloth' in the Bible – for example, Ezekiel 27: 24 describes 'blue clothes and broidered work' being traded by merchants of Sheba and elsewhere in the sixth century BC, but not the origin of the cloth nor the dyes used. When indigo was used in Yemen the source would have been local or imported dye from species of indigofera as woad does not grow there.

Central, southern and eastern Asia

Much of Asia has a humid, tropical climate destructive to organic materials. But this is not the case to the north in central Asia, where, in the Altai Mountains in 1949, was discovered the sensational Pazyryk rug, preserved in ice since its burial in the fourth or third century BC. It is the oldest and most impressive example of pile weaving. Although of unknown provenance, the quality of its weaving and design details suggests an urban context. The wide bands surrounding the central field depict horsemen, deer and griffins in the contrasting crimson of 'Polish' cochineal and soft indigo blue.[27]

In other places religious beliefs and attitudes have discouraged burial and the preservation of used clothing. Both this factor and a seasonal high humidity apply in India, the subcontinent that has exerted the greatest influence on the history of textiles worldwide. Finds from the important Indus valley site of Mohenjo Daro indicate that by the second millennium BC Indian textile technologies were already highly developed. It is probable that indigo-dyeing skills had developed by then too, but early samples of Indian dyed textiles are rare and none exists to confirm the use of indigo.[28] Evidence for its use in India before the medieval period is therefore based on written sources. These include Classical texts like the famous *Periplus of the Erythraean Sea*, written by an anonymous trader in Egypt in the first century AD.[29] As India was the linchpin of early trade both eastwards and westwards, and her people were so highly accomplished in the textile arts, much technical know-how would have filtered along the trading routes, and subsequently her trade textiles were to have an enormous impact globally.

China, too, has always been renowned for the quality and colouring of its textiles, particularly of silk, which were widely traded along the legendary Silk Road linking it to the West from the second century BC. Woven silk fragments from the third millennium BC have been found, and tombs and cave burials have yielded many fine examples from the first millennium BC onwards [30] which reveal a sophisticated variety of textile techniques.[31] There are many specimens of multi-coloured textiles from the Han period (206 BC–AD 220) in the world's museums.[32] A famous concentration of archaeological finds from this period and later comes from the Caves of the Thousand Buddhas at Dunhuang, an important trading centre on the border between Gansu and Xinjiang provinces, where the Silk Road divided into a northern and southern route. The caves were re-discovered by Sir Marc Aurel Stein at the beginning of the twentieth century. Among artefacts recovered were many textile pieces, spanning the Han to Tang (AD 618–906)

Nine fragments found by the archaeologist Marc Aurel Stein in a military fort near Dunhuang in western China. Eight of the fragments, dated between 206 BC and AD 220 come from one large cloth, woven with indigo-blue Chinese designs of dragons, phoenixes and medallions.

dynasties, mostly held now in London (British Museum and V&A) and New Delhi (National Museum). In addition to using dyed threads for embroidery and woven patterns, reserve-dyed techniques such as clamp-, paste-, wax- and tie-resists (see Chapter Six) were practised from the first centuries AD, often dyed with indigo.[33]

Detail from the Pazyryk carpet, fourth or third century BC. Dye analysis has revealed the crimson red to be 'Polish' cochineal (*Porphyrophora polonica* L.) and the blues to be indigo.

(following page)
Tapestry tunic fragment with indigo-blue jaguar. South coast Peru, middle horizon, *c*.AD 700 (35.5 x 21cm).

As for Southeast Asia, Indian *Indigofera tinctoria* seems to have been introduced to the drier parts of the Indonesian archipelago around the end of the first millennium BC, along with the Sanskrit word for indigo, *nila*.[34] There is plenty of evidence for textile production generally in the area earlier than this, as well as trade with southern China and the Indian subcontinent which intensified towards the end of the first millennium BC. Although textile remains are scarce for the early period, a variety of fibres and woven textiles appear to have been in use. By the ninth century AD epigraphic records of Bali and Java, and Chinese trading records, abound with lists of dyestuffs and associated dyeing ingredients. Most of the early references to textiles concern ceremonial cloths whose dominant colours were red and blue.[35]

Central and South America

As in other areas whose indigo production was subsequently exploited by European colonialists on a vast commercial scale, in Central and South America the dyestuff had long been widely used.[36]

For many ancient civilizations woven textiles were an absolutely fundamental part of life. Central and South America, most famously Peru, were no exception. Of paramount importance economically for trade and local use, textiles had great social, political and religious significance.[37] The dyed textiles found in Peruvian graves at Paracas, Nazca, Chancay and elsewhere have, like those of Ancient Egypt, been exceptionally well preserved thanks to the arid coastal climate. The earliest examples are of cotton, others are of cameloid fibres. They reveal that from around

700 BC local cultures had already developed almost every kind of known weaving technique, matched by an advanced knowledge of dyeing. The vivid red obtained from the cochineal insect led the field, but indigo was among the vegetable dyes used on the astonishing woven, embroidered, tie-dyed and otherwise ornamented examples still surviving. As indigo was mysterious, powerful and less common it was often chosen to emphasize prestigious symbols.[38]

Jumping ahead to complete this geographical survey, the later Inca are said to have valued textiles above gold, and sixteenth-century Spanish chroniclers marvelled at the quality of Inca and Aztec weaving and dyeing.[39] Few mountain Inca samples survive, due to the damp climate, but one excavated woman's grave, now in the Museo Amano in Lima, contained a workbox with spindle and balls of dyed cotton badly faded apart from those dyed in strong indigo blue.[40] Unusual post-Conquest Peruvian tapestries include Spanish designs in the traditional coastal colour scheme.[41]

In Central America, too, sixteenth-century Spanish historians noted the local importance of fine textiles as gifts at festivals and weddings and to signify rank.[42] Here, as in Ancient Egypt, blue was generally a revered colour. The Aztecs also used indigo as medicine, and are responsible for the common name in the region for the 'blue herb', *iquilite*.[43] The Mayans mixed indigo with a rare clay mineral to produce 'Maya Blue' paint, widely used for painting on murals, sculptures and ceramics, as well as textiles.[44]

Europe

Archaeological textiles from Europe are fairly rare, as their survival has depended upon special conditions such as peat bog deposits. Enough discoveries have emerged though to show that Europe's neolithic ancestors were producing textiles in the fourth millennium BC, first using vegetable fibres and later wool. Textiles dyed blue with indigo from woad appear in the early Iron Age (*c*.700 BC).[45] Dating from the first centuries AD are many important north European finds such as the famous 'Thorsberg mantle' from Germany, which was checked in shades of blue, and from Denmark garments of blue or blue-and-red check weaves.[46] Woad has also been detected on 'Roman' Iron Age textiles from Polish cemeteries.[47] Although archaeobotanical evidence of woad in northern Europe has existed for some time,[48] it was only in 1992 that evidence from Iron Age England came to light[49] to lend credence to the usual translation of the passage in Caesar's *Gallic Wars* concerning the Ancient Britons' habit of painting their skin blue (see p.223).

Peruvian tapestry (detail), post-Conquest, probably seventeenth century. The colour scheme, including blue highlights on a crimson ground, and weaving techniques are characteristic of coastal Peruvian culture, while designs show European influences.

The Graeco-Roman world

Some Classical archaeological finds and the papyri of Hellenistic Egypt have already been noted, as has the problem of identifying the plant source used to produce early Mediterranean indigo-dyed textiles. The Classical authors, with their passion for recording information, were the first to attempt to understand the origins of the indigo pigment imported from India to the Mediterranean as a luxury item for painting, cosmetic and medicinal use. The *Periplus of the Erythraean Sea* records trade in *indikon* across the Indian Ocean. As indigo was traded in hard stone-like lumps, a common misconception (which persisted in some parts of Europe even into the eighteenth century!) that it was a mineral is understandable. However, in the first century AD Dioscorides, the Greek pharmacologist and botanist,[50] and his (Roman) contemporary Pliny the Elder,[51] apparently realized the vegetable nature of indigo, describing it as froth exuding from 'Indian reeds'. Pliny also mentions a 'second kind' of indigo, the purplish scum floating on a dye vat, presumably of woad or even of shellfish purple.

It can be assumed that in Pliny's day the Romans used native woad for blue dyeing, for with imported indigo pigment at twenty *denarii* a pound, around fifteen times the average daily wage, the several pounds required for a dye vat would have been prohibitive even if they had known how to use it.[52] Pliny, like Vitruvius a century before,[53] notes the use of indigo pigment for fresco paintings, and warns of a counterfeit product made from pigeons' dung or chalk stained with woad but detectable, he writes, with a simple burning test.[54]

Child's tunic, wool with tapestry-woven ornamentation. Egypt, nine–tenth century AD.

By the second century AD the Romans appear to have encouraged the cultivation in their empire of oriental indigo plants as well as woad. There was such flourishing trade with India that maybe technologies, too, were shared. In Roman Palestine, according to Forbes, indigo was known as *indaco* or *kallainos* (a word which may have derived from Sanskrit).[55] This suggests indigofera as the source, for woad was known by its Greek name, *Isatis*.[56] In Egypt the dyeing industry was well organized by the Romans. An excavated Roman dye workshop at Athribis (near Sohag in Upper Egypt, still an indigo centre in the twentieth century) contained sixteen dye vats, most stained blue with woad/indigo.[57] Textiles found at many Roman sites reveal a varied knowledge of techniques and colours, although their provenance is hard to establish. At one, a late first or second century AD Roman desert fortress site on the caravan route linking the Red Sea and the Nile, fragments even include resist-dyed indigo-blue wool.[58] At the Nubian site of Qasr Ibrim, indigo was the main dye used on some exceptional late Roman temple furnishings made of cotton.[59] Many spectacular tapestry-woven and resist-dyed hangings, of wool and of linen, were produced during the overlapping late Roman and early Christian eras.[60] Most of the resist-dyed pieces were found at the famous linen-weaving centre of Akhmim (near today's Sohag), the majority dyed with indigo blue. Concentrated tropical indigo would more easily have dyed hangings of linen and cotton than woad, but the dye source is uncertain.

The Artemis indigo resist-dyed linen hanging (detail). Egypt, fourth century AD (6 x 1.94m, restored from countless fragments).

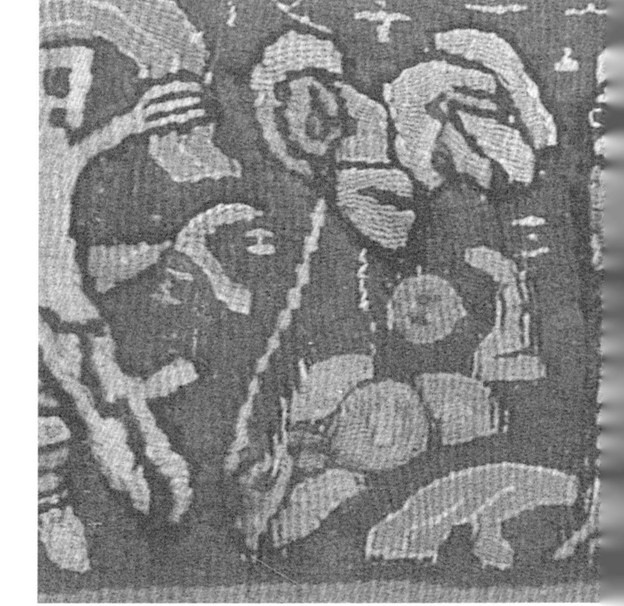

The biblical story of Abraham's sacrifice of Isaac featuring like a cartoon strip in a 'Coptic' tapestry. Egypt, sixth century AD.

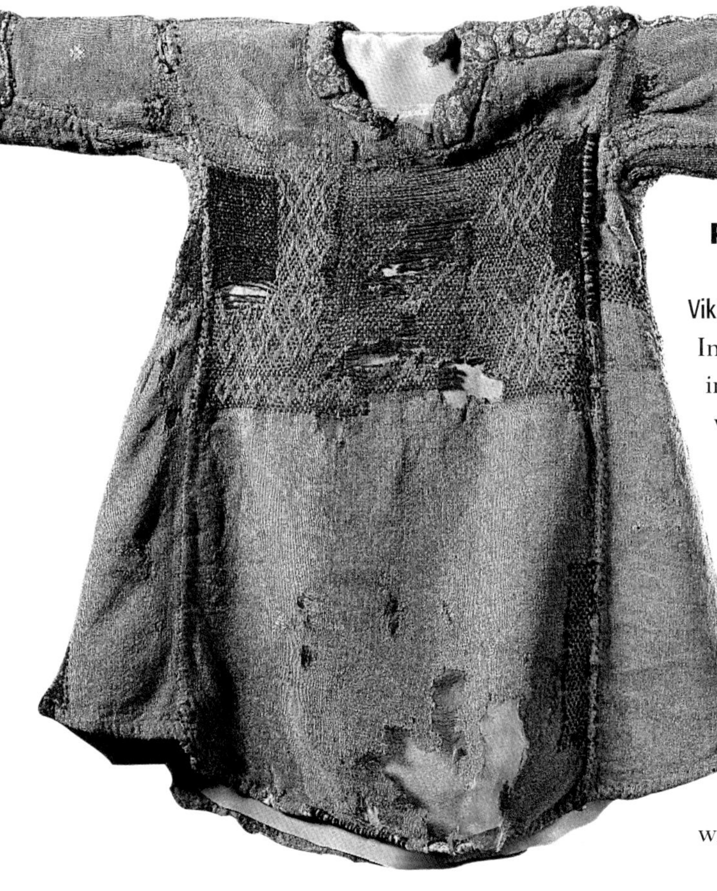

Post-classical and medieval period

Viking textiles

In Europe from the late eighth century the impact of the Viking invasion from Scandinavia was widespread. The Vikings traded in exotic textiles from the Near East to feed Europe's developing taste, as well as introducing their own high quality textiles. Many excavated Viking samples have been analyzed for dyes at York in the north of England, an important Viking centre.[61] It is clear from these that the use of woad was widespread, and that blue was a common colour for clothes for the living, as well as being associated, according to Icelandic sagas, with the goddess of death.[62]

Near East

In the Near East the highlights of the early Christian era are the wonderful Egyptian textiles, mainly dating from the third to ninth centuries, known as 'Coptic' or 'Late Antique' (or, more vaguely, 'Post-Pharaonic'). Earlier examples have been mentioned above, but haphazard nineteenth-century excavations have caused havoc with chronology. Coptic weavers were famed for their skills, and thousands of dyed examples survive, mainly tunics, in which the predominant colours are blues of all shades, reds and purples. For the purples and blacks indigo was combined with madder,[63] and more rarely with kermes, lac or Armenian cochineal, and for green with yellow weld. Shellfish purple, a Coptic dictionary definition being 'genuine purple', is rarely found, so perhaps some of the orders for 'purple cloth' noted in Coptic texts meant the vegetable purple combination. Nevertheless, when the value of money was fluctuating, shellfish purple-dyed wool could be a preferred form of investment, dowry or legacy.

Contemporaneously with 'Coptic' textiles many luxurious silks were being produced in the Near and Middle East, most notably in the Byzantine and Sassanian Empires; in the sixth century the latter stretched from Western Iraq to Central Asia. The few magnificent Sassanian woven silks that survive have mainly indigo in their designs.[64] Dating to the same period is a woollen flat-weave carpet fragment with red and blue woven designs excavated in Persian Khorasan, where the oldest evidence of a Persian pile carpet was also found.[65]

Far East

In the Far East fine dyed silks were produced in abundance, as already noted for China. Indigo was also widely used to dye bast fibres and cotton as cultivation of the latter spread slowly eastwards from the Indus valley. It is thought that the indigo plant and ways to use it probably came to Japan from China by way of Korean artisans together with Buddhism, in about the fifth century AD. The influence of Chinese cosmology with its complex systems of colour coding, and the indigenous Japanese reverence towards the natural world, created a highly developed appreciation of the intricate subtleties and significances of colour in that country.[66] Thousands of textiles of the Nara period (AD 646–794), produced both locally in Japan and in China, have been preserved in the unique Shoso-in repository in Nara thanks to careful packaging in sturdy wooden chests. Indigo features both in these sumptuous Nara woven silk textiles and in resist-dyed cloths which used the three ancient techniques (known collectively as *sankechi*) of wax-resist, clamp-resist and tie-dyeing.[67] In subsequent centuries the use of indigo in Japan increased enormously.

Africa

Sub-Saharan Africa is famous for exotic clothing and weaving, but its climate could hardly be less favourable to the preservation of textiles. Studies by Boser-Sarivaxévanis of the development of textile techniques suggest that knowledge of indigo dyeing, which goes hand-in-hand with cotton production, probably diffused from three main centres, one in the old 'Ghana Empire' (which collapsed in the eleventh century) of Upper Senegal, and the other two in Nigeria – the south-western Yoruba region and (later) the northern Hausa region.[68] But there was almost no concrete evidence of early weaving and clothing until an archaeological expedition to the Dogon region of Mali between 1964 and 1974 made some sensational finds.[69] The Dogon people's unique villages, way of life, myths, carvings and extraordinary festivals have been the subject of much anthropological study. Their predecessors, the now vanished Tellem people, also lived in the strange Bandiagara escarpment, and deposited their dead with their clothing, textiles and other perishable artefacts in burial caves still visible high up the almost vertical cliffs. After each burial the mouths of the caves were re-sealed, thus protecting the contents from the humid

climate. The Dutch team excavated many caves and found around 500 garments and textile fragments dating from the eleventh to sixteenth century AD (when the Tellem appear to have died out), making them the earliest known textiles from sub-Saharan Africa. Most are of cotton, some of wool. The majority of the cotton textiles are coloured in different shades of indigo. The woollen textiles have green, yellow, blue and red decorations, the green being an indigo/yellow combination.[70] This discovery proved that by the eleventh century weaving and indigo dyeing, including tie-and-dye resist, had reached a very high standard in West Africa. Weaving techniques and designs show a marked similarity with those still in use there today, some designated for burials, including many blue-and-white striped patterns, usually weft faced.

Europe and the Orient

Much of indigo's story during medieval times revolves around two pivots. One is Europe's wool and cloth industry with its appetite for woad, the other is the increasing importance of tropical indigo in the East. Both have two main aspects: the production of the dyestuff and its application in the textile industry. For both the scale of trade is awesome.

Until the later sixteenth century imported indigo pigment remained an expensive luxury in northern Europe, largely reserved for paints and inks. All the obstacles and expenses attendant on the long, hazardous journeys by caravan or sea made it an exotic commodity like pepper, other dyestuffs and mordants, medicines and perfumes, which all formed part of the overall spice trade. Across much of Asia, where textile manufacture was the main industry, indigo was a mainstay. As the Islamic empire expanded from the seventh century AD, so, too, did centres of commerce and manufacture, and from the early fifteenth century Ottoman control afforded protection to traders in their empire. Large quantities of textiles and dye materials changed hands in the great markets and ports of the Near and Middle East such as Baghdad, Kabul, Aden, Mosul, Damascus, Aleppo, Beirut, Alexandretta, Fustat/Cairo, Alexandria, Bursa, Antalya and Constantinople. Political unrest affected variously the popularity of the Silk Roads and the more southerly routes. Merchants from many different backgrounds were involved, not least Jews and Muslims (the latter also being redoubtable sailors plying the Indian Ocean). They traded in indigo and dyed textiles from India, the Far East, Afghanistan, Persia, Syria, Palestine, Tunisia and Morocco, while Venetian and Genoese merchants almost monopolized their onward carriage to Sicily, Italy and other ports of southern Europe – Marseilles being a major one. Such goods, being of light weight but high value, were tailor-made for carriage on camel caravans, which could be

Unusual indigo-dyed cotton coif (23 x 19cm), eleventh–twelfth century AD, the earliest known West African textile decorated with the tie-and-dye resist technique. Found in a Tellem burial cave, Dogon area, Mali in the 1960s, where men wear similar coifs today.

made up of thousands of camels. Letters from traders in the Levant describe the eager anticipation when a caravan was due to arrive in a city such as Aleppo. We also learn of the trading impasse caused when camels 'are very skant to be gotten' due to the extreme heat, as an English merchant resident in 'Babylon' in 1583 complains in a letter to London.[71] After such long and difficult journeys, the excitement when oriental wares finally reached the European markets must have been almost tangible.

Prices for indigo dyestuff fluctuated greatly, but were often double those of other luxury goods. Despite the high price, the most sought-after of all types on sale was what European medieval documents refer to as 'Baghdad' indigo.[72] Baghdad was the biggest marketing centre for products made elsewhere in the area, including north-western India and Persian Kirman, whose renowned indigo gave rise to a Persian term *rang-i kirmani*, 'colour of Kirman', for indigo. As early as the mid-thirteenth century dyers of Islamic Spain, thanks to Arab influence, were already *au fait* with the eastern indigo – or 'Indian woad' – vat, which was more concentrated and alkaline than that of woad, and was used for dyeing linens or 'fustians' of mixed cotton and linen.[73] A statute of 1322 shows that woad and indigo vats co-existed in Valencia, but provided that the two dyes should never be combined. As for Italy, no sooner had Italian dyers begun dye experiments with tropical indigo than a Florentine wool corporation's statute of 1317 forbade the practice.[74] However, by the early fifteenth century here, too, the indigo vat for dyeing linens was permitted alongside the woad vat for wools.

Expansion of indigo cultivation

In the medieval Islamic period indigo began to be cultivated on an expanding scale as landowners profited from new ideas and technologies, many probably emanating from India, and from the urban markets. India continued to be a main producer. Marco Polo (possibly at second hand) and others record the main centres as Koulam in the far south and the north-western state of Gujarat,[75] with its port of Cambay being the largest entrepôt for indigo. From the ninth century many Arab geographers refer to the abundance of indigo under cultivation in the Arab world too.[76] Its westward progression seems to have spread from India via Kabul and southern Persia (Kerman and Hormuz were famous for indigo)[77] to the Jordan valley (where it was a major crop until the early nineteenth century), to the oases of Upper Egypt, and across north Africa to the Draa valley in southern Morocco. Berber caravans carried indigo to lands south of the Sahara.[78] Meanwhile, Yemen was always a famous centre, and the Muslims had also introduced indigo cultivation to Cyprus and Sicily, where the dyers were, as was usually the case elsewhere, Jewish.

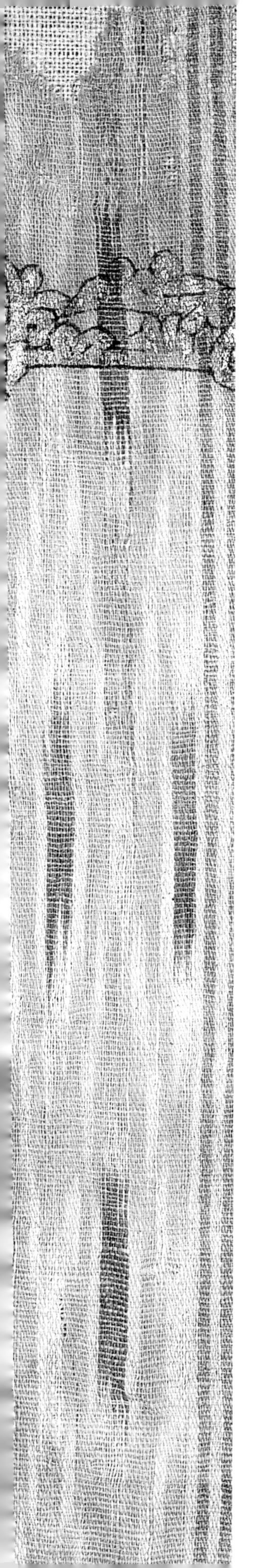

Cotton plainweave ikat fragment dyed in indigo and the yellowish dye known locally as *wars* (*Flemingia grahamiana* Wight and Arn.), with inked Kufic inscription and applied gold leaf. Yemen, tenth century.

Islamic luxury textiles

Under the Islamic caliphates the habit of bestowing textiles as tribute or reward, most notably elaborate costumes featuring inscriptions and known as *tiraz*, stimulated a vast expansion in a textile system inaugurated in Sassanian and Byzantine imperial workshops. Thousands of *tiraz* fragments survive, including unusual Yemen ikat pieces dyed with indigo and *wars*, a popular yellowish dye.[79] Textiles had both enormous significance and value in Islamic life. They were subject to endless nuances of value and meaning, some codified in complex sartorial laws. Trade in Islamic luxury textiles was far-flung, reaching China and western Europe, where many common textile terms derive from Arabic or Persian.[80] European churches filled their treasuries with oriental textiles, including the fine silks for which Byzantium was distinguished until its fall to the Turks in 1453.[81] The European nobility, too, were eager customers for sumptuous foreign silks and brocades. Considering the constant need for dyes to colour all clothing and furnishings, it is not surprising that quality dyestuffs were expensive and that some rulers actively supported their dye-houses. The main dyes apart from indigo were: madder, kermes, brazilwood and, later, lac for reds; saffron, curcuma (turmeric) and safflower (producing also orange and red) for yellows; and gall nuts and sumac as mordants for tannin black.[82] Combinations of these dyes created an infinite variety of colours, indigo featuring especially in greens, blacks and purples. The use of genuine shellfish purple was strictly controlled under Byzantine law. As the *tiraz* system waned, Europe's own textile industry was on the ascendant from the thirteenth century, with an increasing appetite of its own for raw materials, dyestuffs and mordants.

Utilitarian textiles including Indian block-printed trade cloth

It is impossible to tell when dyed clothing became the customary wear of common people, but medieval archaeological textiles provide a clue. Innumerable textile fragments have been excavated, often from rubbish dumps, at Fustat (old Cairo), the Red Sea port of Quseir al-Qadim and other Islamic sites (often on earlier Christian and Roman foundations) in Egypt.[83] An extensive range of textile types and techniques was used, but clearly many fabrics were utilitarian or barter goods, and indigo was commonly used to dye them. Not only are there thousands of fragments of local manufacture, mostly of linen – Egypt's basic fibre at that time – dated before 1100, but there are also many of textiles imported from far away. Of particular interest are a large group of 'Fustat fragments' sold by nineteenth-century dealers in Cairo, which are clearly of Indian origin and should more correctly be called 'Indian block-printed trade cloth'.[84]

Modern block-printed cotton cloth (detail) dyed in synthetic indigo and alizarin red, resembling medieval Indian trade cloths. From Dhamadka, Kutch.

Medieval Indian block-printed resist-dyed cotton trade textile (fragment) with unusual animal designs in light and dark indigo blue. Radiocarbon dated to AD 895 +/- 75 years.

The largest collection is held at the Ashmolean Museum, Oxford.[85] Most fragments are dated between the thirteenth and fifteenth centuries but the earliest, with two shades of indigo blue, has a carbon-14 date of the late ninth or early tenth century. Their provenance is almost certainly north-western India. The fragments are of cotton, mainly resist-printed and dyed in indigo, or mordant-printed and dyed red, or a combination of the two. Many of their designs bear a striking resemblance to those still found on hand-blocked textiles of north-west India.

Woad in Europe

While indigo was on centre stage in the textile industries of the Orient, woad was still firmly in Europe's spotlight. Although for many the word 'woad' conveys hackneyed images of blue-faced Ancient Britons, as a commodity it was of such economic importance throughout the European Middle Ages that it has easily merited a book of its own, Jamieson Hurry's 1930s' classic *The Woad Plant and its Dye*.[86] Its main predecessor, with its well-known plates (see pp.56 and 93), is the German monograph of 1752 by Daniel Schreber.[87]

Woad, as already mentioned, was Europe's truly universal dye, used for all blues but also as a 'top' or 'bottom' dye for most other colours.[88] As one Elizabethan put it: 'No colour in broadcloth or kersey will well be made to endure without woad,'[89] and a textile historian has noted that 'in medieval draperies, woad-indigo served as the essential foundation . . . for creating a wide range of other colours' (including so-called 'scarlets').[90] The secrets of the medieval dyer, recorded in many documents and guild regulations, are increasingly being confirmed and extended by chemical analysis. Such was his skill that he was able to reproduce small nuances of hue. Beautiful greens, known in Britain as Saxon, Lincoln and Kendal greens, were made by combining woad with yellow from weld or from dyer's broom (*Genista tinctoria* L.) – hence the plants' common names, 'dyer's greenweed', 'dyers' woad' and 'woadwaxen'. A further yellow widely used on the continent came from saw-wort (*Serratula tinctoria* L.). Purples, pinks, violets, mulberries and scarlets came from woad combined with madder or kermes reds, or even lichen purple.[91] It was needed, too, to make russet, tawny and grey colours, as well as the best blacks. The highly coveted black known in the Mediterranean textile centres as *bruneta* (brownish) black was made from woad and madder. To ensure the blue shade was dark enough before proceeding to the madder stage, the quality was strictly regulated by guild statutes. Generally forbidden, especially for dyeing fine wool, were inferior *negre* blacks made from corrosive plant tannin/iron mordant mixtures.[92] Finally, woad alone could, under certain dyeing conditions, yield greens or even, with the addition of alum, pinks.[93]

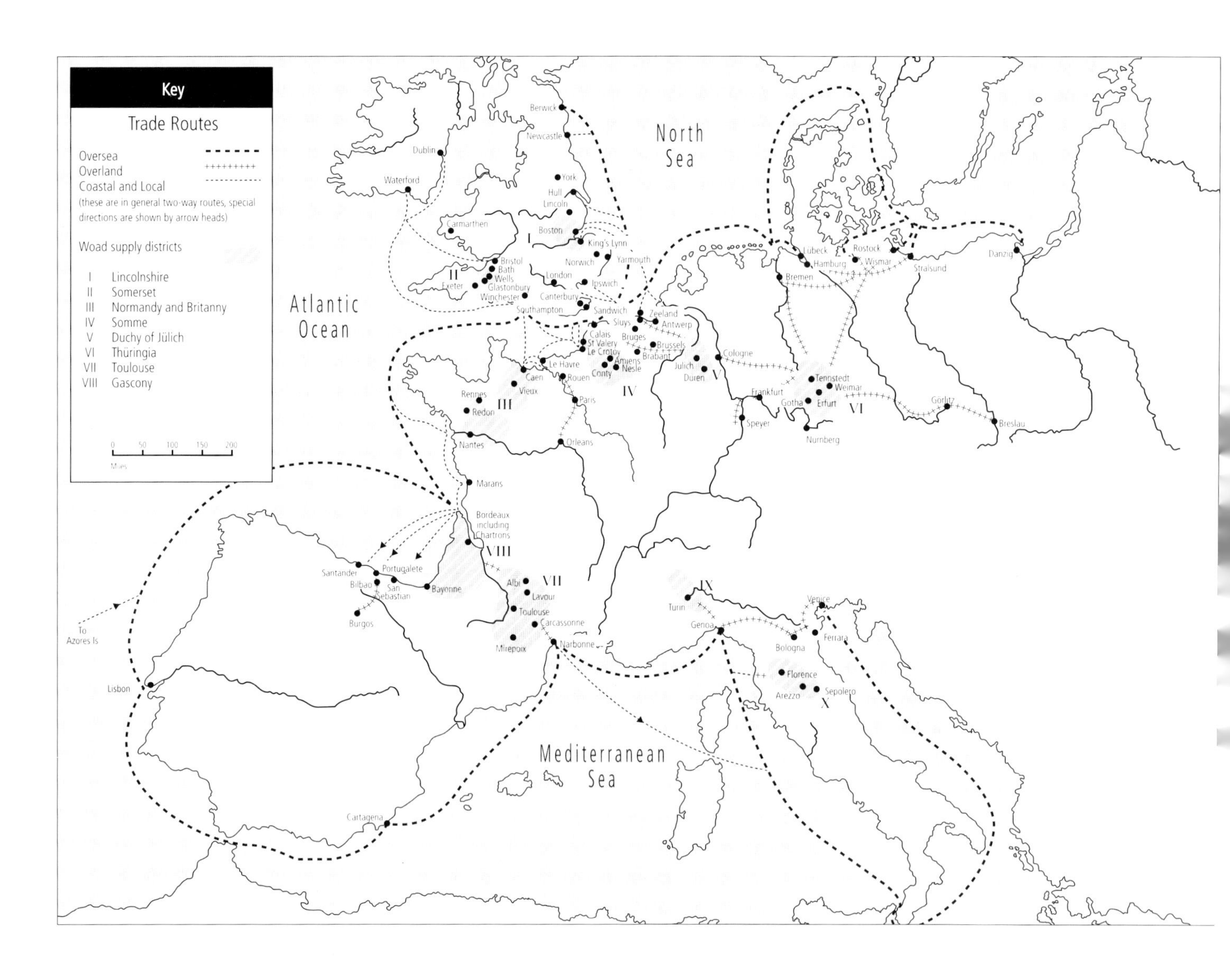

Map showing the trade routes of woad and chief districts of production in medieval Europe.

[opposite above]

Woad merchants with a bulging sack of woad balls. South exterior wall of Notre-Dame cathedral nave, Amiens, France. Donated by the woad merchants' corporation in honour of St Nicholas, regarded as their patron saint, c.1300.

It is not known who first began to cultivate woad, but by the thirteenth century in western Europe the livelihoods of many farmers, merchants and dyers revolved around its production. Governments also gained greatly by imposing heavy taxes at every stage. A poem by Chaucer, harking back to a pastoral age unspoilt by greed and industrialism, laments the harmful dyeing industry with its three staples, madder, weld and woad:

No mader, welde, or woode [woad]
no litestere [dyer]
Ne knew; the flees was of his former hewe.[94]

Regions most renowned for woad production were in France, Germany, Italy and, later, in England.[95] Spain's central regions grew woad,[96] but a good deal more had to be imported. Much woad trade was handled by Spanish merchants, particularly from Burgos,[97] and imported tropical indigo was also available in the Muslim south.

France was Europe's greatest producer of woad, called there pastel or *guède* (or, further north, *vouède*).[98] In addition to local consumption, exports, particularly to Britain and Flanders, were huge, although political events like the Hundred Years' War caused fluctuations. The main woad districts were Normandy, Picardy and Languedoc.[99] Although yields were lower in the north, cultivation and trade was still extensive, notably around Caen, Bayeux and Amiens. Late thirteenth- and early fourteenth-century city records and customs accounts show that Amiens merchants traded enthusiastically in Picardy woad. The industry was tightly regulated. A typical by-law of even the small town of Corbie, near Amiens, admonishes:

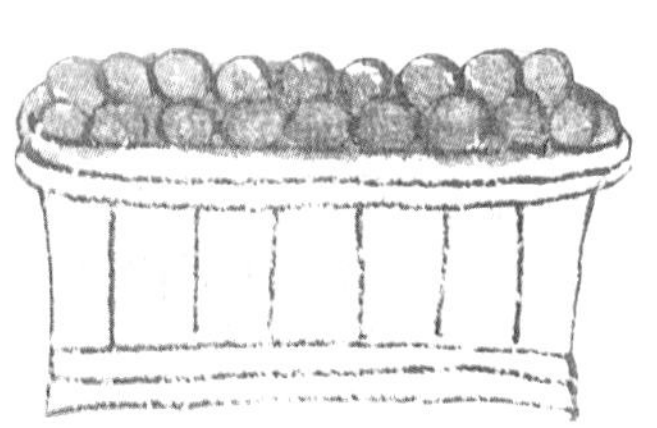

Woad balls in a basket from the margin of the land book of the Archbishop of Cambrai, which lists agricultural commodities and tax dues, dated August 1275.

No merchant shall buy or sell woad at the market of Corbie or in the outskirts of the town until the sacrament bell has rung, nor on other days until the collector of tolls or his agent shall have struck the beam and thus given the signal.[100]

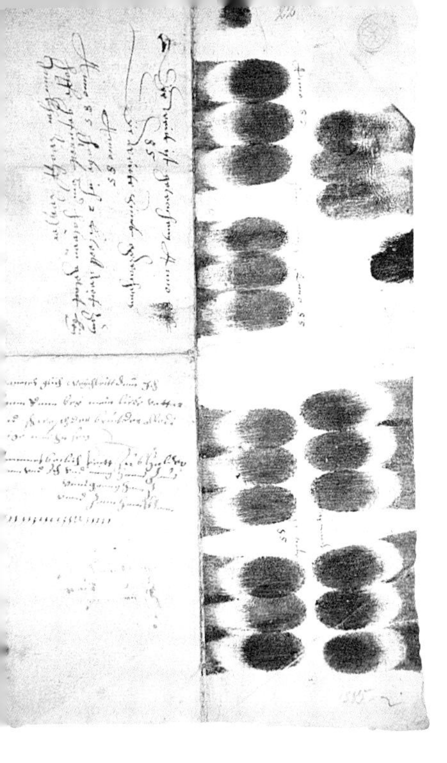

Page of woad samples from Erfurt, Thuringia, Germany, 1585. Prospective purchasers gathered in the market to test woad qualities by passing damp fingers over woad balls and smearing them onto sheets of paper.

It was in the Languedoc region of southern France, renowned for its woollen cloth industry, that woad reigned supreme. One writer on Languedoc commented that 'woad . . . hath made that country the happiest and richest in Europe'.[101] Its appellation as the *pays de cocagne* – land of the woad ball – even came to mean exceptionally profitable land anywhere. The prosperous woad merchants, particularly of Albi and Toulouse, were often landowners too, overseeing every aspect of woad production and trade.[102] They displayed their affluence in opulent houses, many of which still stand today in the old centre of Toulouse. One merchant, the immigrant Castillian Jean de Bernuy, was so credit-worthy that he could even stand as the main guarantor of the sum required for the ransom for King Francis I after his capture by Charles V of Spain.

France's southern cloth industry consumed much Languedoc woad, but large quantities of it were transported overland to major ports like Bayonne and Narbonne. The rest travelled by barge along the Garonne from Toulouse to Bordeaux, which profited greatly from its export, often on Italian vessels, to Flanders, the Low Countries, Italy and above all to Spain and Britain, both in turn major exporters of wool and cloth to Europe.[103] Major goods exported alongside the annual 200,000 bales of woad were madder dye, alum mordant and, of course, wine. Such was the importance of woad that even in times of war unarmed ships were permitted to enter port to load it.

In Germany the economic prosperity of the districts of Jülich and Thuringia was also founded largely on the woad industry.[104] In its heyday woad provided a third of the income of Thuringia (the other noted products being wine, hops and wheat). Most visible relics of the local industry of at least 300 villages are the quantities of woad mill stones still scattered around the countryside.[105] The wealth woad generated is largely responsible for the beautiful medieval city of Erfurt, including its university, at the heart of Thuringia's five *Waidstädte*, or woad towns – the others being Gotha, Langensalza, Ternstedt and Arnstadt. Prosperous woad merchants proudly called themselves *Waidherrn*, 'gentlemen of woad'.[106] In Erfurt today you can pass their grand houses whose wide entrances were constructed to accommodate large wagons, and stroll down the Waadgasse, where woad was weighed and 'assayed', i.e. officially tested. In churches in the city a carved keystone, a stained glass window (see p.10), and a tombstone all commemorate the woad industry.

Land register (detail) of 1640 showing a woad mill in the centre near a village church. Erfurt district, Germany.

The German woad industry was strictly regulated at every stage and heavily taxed. As early as 1250 duties known as 'woad pennies' were being levied at Erfurt, and companies were formed to provide the capital necessary to trade in woad. Particularly complex ordinances directed the farmers, 'refiners' and merchants who handled 'green' (unfermented) and 'couched' (fermented) woad.[107] Under the protection and organization of the Hanseatic League, formed in the thirteenth century, German merchants, as well as French, exported their woad to many parts of Europe, including Antwerp, England, Hungary, Poland and Italy.

Italy's flourishing textile industry, especially in Florence, had such an insatiable demand for woad, called there *guado*, that home production had to be supplemented from France and Germany, hence her early adoption of oriental indigo imported by Venetian merchants. However, several regions profited greatly from woad cultivation, notably Tuscany,

Lombardy and Piedmont, and many large millstones have been preserved.[108] As the Florentine cloth industry was renowned for its high standards, much attention was paid to quality control at all stages, not least for woad, the major dyestuff. It had to be stored in a great central woad hall where it could be thoroughly inspected and costed before sale.[109]

Italian merchants, centred on Venice, dominated much of the European maritime trade in woad, just as they did the oriental spice trade, which included indigo pigment, in the eastern Mediterranean. The principal route from Venice went via Genoa, Narbonne and Bordeaux to Britain, the Low Countries, and north-west Europe. In the fourteenth century much woad, and ash for the vats, was exchanged by Genoese merchants in England's south coast port of Southampton for wool and woollen cloth, for at this period Britain's own woad production was not all that significant.

In southern England woad was being cultivated from Anglo-Saxon times, as many names with the word 'wad' in them testify;[110] but Britain nevertheless relied greatly on imported dyestuff until the later sixteenth century, especially as her textile industry developed. Before the fourteenth century vast quantities of English wool and undyed cloth were sent across the channel to be dyed, woven and finished on the continent by artisans, notably Flemish, with superior skills. From 1330 Edward III encouraged Flemish dyers, fullers and weavers to bring their expertise to England; the English cloth industry improved and expanded greatly and with it the need for dyes and mordants.[111]

As elsewhere, shipping manifests, port records, city ordinances and other documents paint a vivid picture of woad's importance in the textile industry.[112] It was so valuable, second only to wine, that it was often bequeathed as property in wills.[113] Between the thirteenth and the sixteenth centuries in particular it was in constant demand, along with allied dyeing materials and linen. Its trade brought great wealth and status to the merchants who monopolized it.[114] A city like Winchester, a staple town strategically placed between the wool-producing regions of southern England and the key ports of London and Southampton, was a major trading and industrial centre. In a fascinating account of its dyeing industry in *Survey of Medieval Winchester* the author states:

Of all the cloth-manufacturing trades that of the dyer . . . was probably the one which required the greatest capital investment and the highest degree of entrepreneurial expertise. Before the later fourteenth century the Winchester dyers appear to have been the wealthiest of the cloth-working craftsmen.[115]

Dyer's letter of February 1626 to woad-growing landowner and lawyer Henry Sherfield (Member of Parliament for Salisbury). The letter, to which is attached a dyed wool sample, begins: 'Syr I have sett your woad it is a very kynd woad to work but the value is nott to be sett of it by eas.'

Like the wool merchants, dyers held prominent civic positions. A church near the dye market was even known as St Mary Wode. Court rolls of the time reveal what sound like a modern preoccupation with environmental pollution, for dyers were regularly 'presented' for contaminating the city's water supply with 'wodegore' (woad dye waste).

Winchester had to compete with many other cities to secure woad supplies.[116] Bristol, for example, was one of England's main woad ports and cloth centres, and her book of records of 1344 to 1574 includes ordinances to control the marketing of woad and to protect skilled dyers from rogue usurpers:[117]

Divers men as well as those who had not been apprentices, servants or masters of the said craft . . . not having knowledge of the aforesaid craft of dyeing have taken on themselves to dye cloths and woollens put in woad . . . the which cloths because of bad management and through lack of knowledge of the said men are greatly impaired of their colours and have many other defaults to the great loss and damage of the owners of the said cloths and to the great scandal of the town and shame of all the craft aforesaid . . . Wherefore it is ordained that henceforth no man of the said craft nor of any other craft undertake to dye any cloth or wool unless it be presented by the said masters that he is good, and able and sufficiently learned in the said craft . . .[118]

In the second half of the sixteenth century Britain's textile industry received a boost with new influxes of foreign artisans, including Dutch refugees, bringing their expertise. In the same period the price of woad and other dyestuffs, particularly from Toulouse, was becoming inflated, and religious wars on the continent also threatened imports. These factors worried the government and other interested parties, and stimulated the local production of dye materials as well as the quest for alternative sources of supply.[119] The famous geographer Richard Hakluyt, in a note of instruction dated 1578, tells his explorers: 'If we may enjoy any large territorie of apt soyle, we might so use the matter as we should not depend upon . . . France for woad, baysalt and gascoyne wines . . . so we should not exhaust our treasure and so exceedingly inrich our doubtful friends, as we do.'[120] To bypass her rivals, Britain turned to the islands of the Azores (which profited much from woad exports between 1550 and 1650)[121] and the Canaries for new sources of foreign woad.[122] An employee of Sir Walter Raleigh even explored the possibilities of cultivating woad in the new lands of Virginia:

Woad: a thing of so great vent and uses amongst English diers, which can not be yeelded sufficiently in our owne countrey for spare of grounde, may be planted in Virginia, there being ground enough. The growth thereof neede not to be doubted, when as in the Islands of Asores it groweth plentifully, which is in the same climate. So likewise of madder.[123]

As for home production, woad rapidly became an established alternative crop in southern Britain. Its cultivation spread along the south coast and inland up the fertile river valleys as far north as Scotland.[124] In the government census of woad growing in 1585, the county of Hampshire, with 1,748 acres, grew the most. In Somerset, a county famous for cloth, woad cultivation was a prime source of wealth for the abbeys, notably Glastonbury, whose name may even derive from the Latin *glastum*, taken to mean 'woad'. The fenlands of Lincolnshire, East Anglia and Cambridgeshire were also important centres.[125] In the mid-1580s vital grain supplies were threatened by the craze for woad growing – said to be six times more profitable – which had to be restricted by government licence. But such was the resentment this caused that Queen Elizabeth I lifted the restriction in 1601, on condition that no woad processing would take place close enough to her palaces to offend the royal nose. Archival evidence paints a lively picture of the woad industry.[126]

Despite the wide-scale availability of imported oriental indigo from the seventeenth century, woad was still grown and used in Europe, and all those with a vested interest fought hard for its survival. As well as being a lucrative crop for the producers,[127] its labour-intensive processing provided valuable work for the unemployed. The production of woad in parts of Europe, therefore, survived on a surprisingly large scale throughout the seventeenth and eighteenth centuries. Even in the twentieth century, woad was still being added to indigo vats, mainly for dyeing woollen cloth for service uniforms, and for the same reason considerable amounts continued to be exported to the United States.[128]

Ironically, Britain, the first northern European country fully to embrace imported indigo, was the last to abandon woad, for the final crop was processed in the Lincolnshire fens in 1932.[129] Despite this, little concrete evidence of the industry remains, as most English woad mills were temporary structures operated by itinerant workers and after 1880 used wooden mill wheels. Elsewhere Rhind records the ongoing cultivation of woad in the nineteenth century, not only in France and Germany but also in the Azores, the Canary Islands, Italy, Switzerland, Sweden, Spain and Portugal.[130] Others mention Hungary, too.[131] Generally, however, the woad industry was struggling, and occasional deliberate attempts to resuscitate it, not least in France during the continental blockade (1806–12) when Napoleon was desperately seeking home-grown alternatives to imported indigo, were not a success.[132] France gathered in her final woad harvest in 1887, while the woad mill of Pferdingsleben in Thuringia, Europe's only one still *in situ*, finally ground to a halt in 1912.

The stage had been set for woad's relegation to the bottom division of the dyeing league, defeated by indigo's meteoric rise ever since the momentous rounding of the Cape of Good Hope by Portuguese ships in 1498.

Woad mill in Pferdingsleben, Thuringia, Germany, photographed in 1910. The disused mill still stands in the centre of the village.

Not a chest of indigo reached England
without being stained with human blood
(E. DE-LATOUR, 1848)

Indigo's Heyday

The Downfall of Woad and Salvation by Denim

'The search for export commodities such as indigo, saltpetre, raw silk, coffee, and tea took all the European companies from one end of the Indian Ocean to the other.' So writes the historian Chaudhuri in one of his fascinating examinations of the subject.[1] There is only room in this study to skim the surface of the contribution of indigo to four centuries of turbulent trading history.

Vasco da Gama's circumnavigation of the Cape of Good Hope in 1498 marked a crucial turning point in indigo's fortunes, as did the Spanish conquest of America in the following century. The Portuguese achievement made possible direct importation by Europe of goods by sea from India, the Spice Islands, China and Japan, and thus avoidance of the heavy duties levied by successive rulers on Asian goods whether in transit overland to the West or by the old sea routes around the Arabian peninsula. In this way the Portuguese broke the Middle Eastern and Italian commercial control over trade in spices, which included dyestuffs, and in luxury textiles; and the new competition led to a gradual decline in the wealth of Venice and a changing pattern of trade in the great Arab centres. From the early seventeenth century the Portuguese were followed by the Dutch, English, French and Spanish, who also established trading bases in the area. The entry of these other Europeans into Asian trade enormously increased supplies of Eastern goods reaching Western markets, and lowered their

Sixteenth-century pew end in Bishops Lydeard church, Somerset. Sailing ships feature in pew ends and stone carvings in West Country churches, reflecting the nation's seafaring and trading history.

prices. Indigo ranked among the major commodities carried on the new powerful 'East Indiamen' sailing vessels. It was often the most valuable of all the 'spices', and progressively undermined the whole European woad industry. The largest emporia were Amsterdam and London, where during the seventeenth century a vast re-export trade developed in calico and spices, including pepper, cloves and indigo.[2]

Thereafter the fate of indigo see-sawed between East and West. It was always a commodity that attracted speculators and fuelled rabid competition. Profit margins fluctuated greatly, often due to over-production. In any case the commodities listed above by Chaudhuri were all highly sensitive to price changes.[3] Although indigo from the East began to outweigh woad, it was itself out-balanced from the middle of the seventeenth century by the development of indigo plantations in European colonies in the West Indies and the Americas. But economic and political factors, notably the American Revolution and the Napoleonic wars, would again send the see-saw lurching back in favour of India and Java, the former witnessing a dramatic expansion in indigo production in the nineteenth century. And later, just as the availability of synthetic indigo was undermining the Indian industry at the beginning of the twentieth century, the First World War, strangling the German factory product, temporarily tipped the balance back again in favour of natural Indian indigo, thus giving the planters a final short-lived reprieve. The war over, synthetic indigo flooded the market and commercial production of natural indigo slid rapidly into the ground, although small amounts of Indian and Mexican indigo were still appearing on the European market for a long time after the Second World War.

For four centuries the many-stranded story of indigo, as of other commodities much in demand, is woven into an immensely complex pattern of trans-continental trade, colonial agricultural enterprise based on slavery, revolutions in industry, rivalry between the great European shipping companies and political relations between trading nations. A genuine passion in the West, and elsewhere, for beautiful textiles and exotic dyes and spices was fuelled by greed, slavery and protectionism. It would be unjust though to ignore the pioneering enterprise and the genuine spirit of adventure which lay behind it all. Today's global trade battles and rival political ambitions, still aggravated by widespread exploitation of human and environmental resources – but lacking the old stimulus of adventure – impart a depressing sense of *déjà vu*.

Hand-woven linen towels from Perugia, Italy, with supplementary-weft patterning in indigo-dyed cotton. Sixteenth century.

The East India companies

In 1510, twelve years after landing in south-west India, the Portuguese reached the Spice Islands of the East Indies. For the rest of that century they, the Spanish, the English, the Dutch and, to a lesser extent, the French rivalled each other to corner the spice trade, above all in pepper. Portuguese traders held pride of place in the sixteenth century, and still retained a large portion of trade with west India in the first decades of the seventeenth. A study of Portuguese trade with Asia between 1580 and 1640 particularly emphasizes the importance of indigo throughout this period.[4] The spice trade was officially a royal monopoly, but from the 1550s royal control weakened and much private cargo found a place in the king's ships. This was largely handled by 'New Christians' (i.e. forcibly converted Jews) who outwitted the king's attempts to enforce quotas. It remained the case that only the king's agents could carry pepper, but private merchants made up the cargoes with goods of high value relative to bulk. This was indigo's forte. Although textiles outranked all other items in quantity (and we cannot tell how much indigo was present in these), *drogas* (i.e. 'spices', including dyes) averaged 22 per cent of ships' cargo between 1580 and 1640. Despite high royal customs duties imposed upon it, indigo was still easily the most valuable of the *drogas*, generating consistently attractive profit, 'considerably more than that of the King's pepper, though only one-sixth of its volume'.[5] For this reason indigo sometimes formed as much as three-quarters of the total private shipment of *drogas*. In the 1580s, despite persecution by the Inquisition, 'New Christian' merchants managed to dominate the indigo exports of Cambay and Goa, to the annoyance of the king and the 'Old Christians'. One's imagination is fired by the picture of maritime and coastal activity as fleets of sailing vessels of all kinds to-ed and fro-ed up and down the coast of western India, arriving at Goa, Portugal's main entrepôt, piled high with exotic goods which included fine textiles, indigo, opium, camphor, pepper, gilded and lacquered bedsteads, amber, crystal, gold and precious stones. By the 1620s Portuguese shipments of cloth and indigo to Europe were declining, faced by growing Dutch and English competition and saturation of the market.

The Dutch companies, having merged to form the powerful Dutch East India Company (VOC) in 1602, had gained the upper hand in the Spice Islands by 1619 and established Batavia in Java as the capital of their East Indian empire. Largely by using as barter Indian painted cottons, already the principal currency of exchange, the Dutch brought spices, cotton, silks, tea and other goods, including Javan and Indian indigo, to Europe. Indigo cargoes alone became massive: according to one source, seven Dutch ships in 1631 carried between them a total of 333,545 pounds of indigo worth at least five *tons* of gold.[6] Dutch dyers were renowned for their skills; hence the term 'Hollands' which referred to undyed cloth sent to Holland for dyeing. In the early seventeenth century when other European nations were, as we shall see, busy resisting the foreign dye that clearly threatened their woad industry, Dutch dyers had a distinct advantage. For the rest of the century more and more Java indigo arrived in Holland, both for home use and increasingly for distribution throughout Europe.[7]

England's East India Company (EIC) was founded in 1600. Although the Company left the Spice Islands to the Dutch, it struggled for half a century to wrest from them, and from French and Portuguese contestants, supremacy over sea trade with north-west India. As well as seeking markets for her broadcloth and supplies of spices, the EIC wanted to undermine the indigo trade of rival Levant companies, who used the overland routes and were, with the Portuguese, Europe's main suppliers of indigo at the turn of the seventeenth century.[8] Indigo had long been established as a major export from north-west India, as Portuguese chroniclers such as Pirés[9] and Barbosa reported in the early years of the sixteenth century. The latter listed 'fine indigo in cakes and another coarser kind' as among the products traded by 'the great and wealthy Kingdom of Guzerate and Cambaye'.[10] Hakluyt had recorded the suppliers of Indian indigo in the 1580s as being 'Zindi [Sind] and Cambaia'.[11] For at least a decade after the EIC established commercial relations with Surat in 1608, indigo was by far the most valuable item of its Indian trade, followed by calico, saltpetre, sugar, cotton yarn, cotton wool, ginger and gumlac.[12] In the 1630s, however, the situation changed, partly due to the disastrous famine in western India, and the sale of calicoes began to overtake that of indigo.

In the words of the textile scholar Gittinger: 'It has been judged that there are nearly three hundred dye-yielding plants in India. Of these, none was both artistically and commercially more important than indigo.'[13] The Indians had for centuries been processing this dye by traditional methods.[14] In the north the Dutch were soon exploring a good source of indigo (traded in the form of cakes known in England as 'flat' or 'round') around Sarkhej, near Cambay, a fact later noted by English and French travellers and traders.[15] Many nearby towns also traded in indigo, as did places in Sind. The few surviving European merchants' houses in coastal towns of Gujarat stand testimony to a fascinating period in early colonial maritime trade. Inland the Bayana district, near Agra the Mughal capital, was the source of the purest indigo, known as 'rich', or 'Lahori'/'Lauri', the latter name reflecting the long-established importance of Lahore as an entrepôt for indigo going west overland.[16] It has been calculated that in its heyday indigo was cultivated on well over 10 per cent of the total arable land around Bayana.[17] (As late as 1901, the area of Agra and Oude still had around 150,000 acres of indigo under cultivation.)[18] Pelsaert was commissioned by the VOC in the early seventeenth century to provide a detailed report on Indian indigo production,[19] and the Frenchman Tavernier soon followed suit.[20] At the end of the century the VOC called for a memorandum on Sarkej indigo, since they were interested in expanding indigo cultivation in Indonesia.[21]

From the first decade of the seventeenth century competition between rival European, Armenian, Persian and Indian traders intent on securing the best supplies of indigo was frenzied. Pelsaert paints a vivid picture of traders' wily ways,[22] as do English traders' reports quoted in Foster's *Early Travels in India*. When the merchant William Finch, for example, was dispatched by the EIC in 1610 to buy Bayana indigo he even created a diplomatic incident by daring to outbid the emperor's mother's own agent, thus antagonizing the emperor himself. Unabashed, Finch took his 'twelve carts laden with *nil*' to Lahore, and thence, after an acrimonious dispute with his distrustful colleagues, decided to go on via the traditional overland route to Aleppo. In fact, Finch and his party died in Baghdad, causing further feuding over the indigo.[23] Many similar anecdotes illustrate the lengths to which the early traders would go to secure a commodity as precious as good indigo. The obstacles to be overcome in securing and transporting it certainly sound formidable. As well as illness and the hostility of local chiefs, dangers on land included 'theevish beastly men… savage beasts, lions, tygres etc.' and even 'such infinite numbers of munkeyes, leaping from house to house, that they doe much mischeife and, untyling the houses, are readie to braine men as they passe in the streets with the stones that fall'. By sea, in addition to frightening storms and the constant threat of attack from foreign vessels and pirates, the Bay of Cambay harboured special perils,

being dangerous to passe by reason of the great bore which drownes many, and therefore requires guides skilfull of the tydes . . . Theeves also, when you are over the channell, are not a little dangerous, forcing you…to quit your goods, or in long bickerings betraying you to the tydes fury, which comes so swift that ten to one you escape not.[24]

In India the marketing of indigo was complicated, speculation was rife and prices were volatile. Major traders tended not to use the spot market. Wealthy local merchants purchased the best crops in advance, leaving other traders to buy on the spot for cash or negotiate advance orders for the following season.[25] Purchasers had to be on the lookout for fraud, as producers added substances such as sand, oil, wood shavings or moisture, or used other tricks including dubious packaging methods, to increase the volume and weight of the product.[26] Not that traders in Europe were above practising fraud themselves – Court minutes of the EIC warn against those who buy cheap 'dust' of indigo to fashion into cakes and palm off as good quality indigo. This, though, could have been jealousy, as the Company was furious when an enterprising grocer was granted a patent in 1634 to refine and purify cheap indigo dust.[27]

Quality and price of indigo and other export goods were overseen by being channelled through Company depots known as 'factories', the most important in north-west India being established at Surat in 1608.[28] Prices and profits constantly fluctuated over the years to keep pace with supply and demand,[29] and in London on at least one occasion dividends for 'adventurers' (as EIC shareholders were called) were paid in indigo as well as money. Unforeseeable factors affected the market. If, for example, one of the many ships that went down at sea had 1000 barrels of indigo on board, it was a financial disaster. In the 1620s the Court of Directors of the EIC would keep selling prices artificially low to undercut Levant trade or hold stocks back until the spring season for cloth dyeing. Around 300,000–400,000 pounds weight (i.e. about 160,000 kilos) would be sold per year, the best for over six shillings the pound, huge sums in those days. It was sold variously in barrels, 'bales', 'fardles', 'shirts' (calico wrappings), 'skins' (bags or chests bound in leather) and as 'dust'. After the mid-1620s the re-export market took off in a big way, with eastern indigo being re-exported from Britain to Amsterdam, Mediterranean ports and the Levant. However, from around 1640 the EIC had to reduce supplies from India in face of competition from the West Indies and Spanish America. The Company therefore constantly exhorted its agents in India to ensure the quality was good and prices low, as well as trying to prevent private traders from undercutting its own investment.[30] Despite the competition, indigo trade, most via the major port of Bombay which had been acquired from the Portuguese in 1661, was still considerable.[31] EIC Court minutes of the period often refer to the appointment of 'Keeper of the Blue Warehouse' in London.[32] Indigo supplies were greatly affected by political factors, especially wars at home and abroad. Sea battles could cause rapid price rises or shortages, especially when West Indian trade

Original page listing a general Court of Sales of the East India Company, 3 December 1651. Included in the list is Sinda indigo, and indigo shirts and skins from Lahore and Sarkhej.

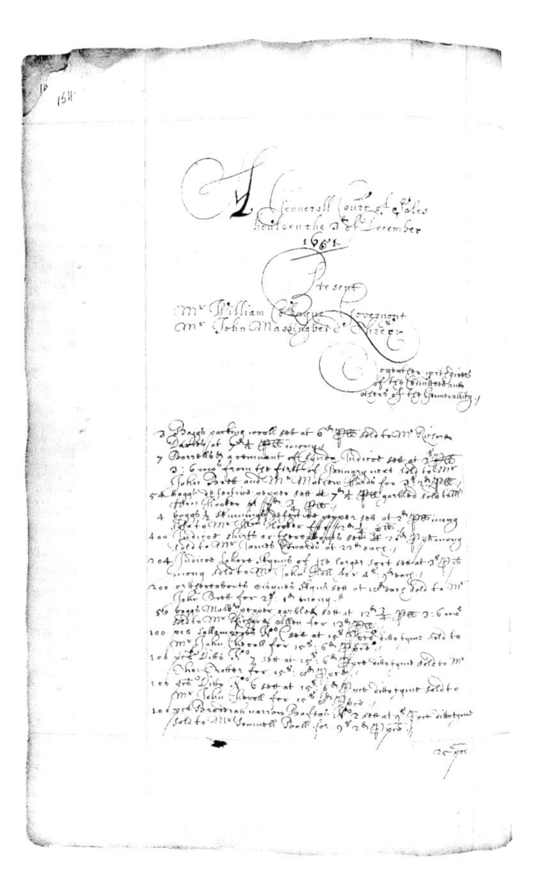

was disrupted. In 1695, for example, a French ship laden with indigo was sunk by an English man-of-war, much to the delight of the EIC Court, who soon plugged the gap with Indian indigo.[33] After the 1720s, indigo imports from India were spasmodic, but there were still some good years when it was as much as 10 per cent of total EIC imports from Bombay.[34]

From the middle of the seventeenth century, southern and eastern India had gained in importance for the European companies, based on ports along the Coromandel coast, a long-established Arab and Persian trading region. The Dutch and English were followed there by the Danish, whose company was founded in 1612, and the French, although their company was not founded until 1664. Although much indigo dyestuff was manufactured and traded, the most vital exports from the region were exquisite cotton and silk textiles of all kinds, above all the famous painted 'pintadoes', or chintzes. Plain cloths included many 'blue calicoes'. The main attraction of all Indian dyed textiles was their colourfast and brilliant colours.[35] These included, as well as the mordant dyes, incalculable amounts of indigo. We cannot therefore limit European imports of indigo simply to the raw dyestuff.

From 1600 to 1800 India was the greatest exporter of textiles the world has ever known.[36] At first they were needed by the Europeans, as they had been by their Arab predecessors, for bartering in the Spice Islands, but soon they were also being re-exported from Europe to North Africa, Turkey, the Levant and West Africa (as barter for slaves). From the middle of the seventeenth century, despite various attempts to ban them, they became all the rage in Europe to the dismay of the home textile industry. Chaudhuri suggests that the popularity of imported cotton textiles was a root cause of the Europe's Industrial Revolution, which started with the mechanization of spinning and weaving.[37] Steps taken to learn Indian methods of dyeing patterned cloths (see pp.158–9) were also crucial to its success.

From the 1630s the Dutch and British were also exporting from trading posts in Bengal quantities of raw silk, luxurious muslins and other textiles, and saltpetre. Indigo dyestuff was a fairly minor export from Bengal at that time, but as usual one cannot determine how much indigo was exported 'invisibly' in dyed textiles such as 'blue ginghams' that appear on trading lists.[38] In 1691 the British were permitted to settle in an obscure swampy village called Calcutta, which in the eighteenth century they transformed into a thriving city. Bengal was to become the world's greatest indigo supplier, the wealth generated by the British indigo trade adding greatly to Calcutta's prosperity. But before then other parts of the world, as we shall see, had their day.

Painted and dyed Indian cotton floorspread (detail). From Golconda State, made for the Persian or Indo-Persian market, *c.*1630 (325 x 246cm).

The Levant companies

The well-established commerce in valuable dyestuffs, spices and textiles in and through the Middle East until the second half of the sixteenth century was described in the last chapter. Until goods destined for Europe reached Italian middlemen in the Mediterranean, this trade was largely handled by local merchants. The domination by the Portuguese of the Cape route to the East in the sixteenth century stimulated other European merchants to seek access to the source of Asian goods by alternative routes which we shall call 'Levantine'. These, starting from the eastern Mediterranean, either went all the way east by land through Mesopotamia, Persia, southern Russia and Afghanistan, or diverged down the Red Sea and the Persian Gulf into the Indian Ocean. The use of these Levantine routes, whether by Middle Easterners or Europeans, continued right up until the beginning of the twentieth century. Baghdad remained the largest inland entrepôt for goods from the East, while Cairo was still a main recipient of those shipped around the south Arabian coasts. All Middle Eastern markets were avid for indigo, described in 1801 as being 'of foremost necessity in the whole of Asia'.[39] The passage of huge quantities through Aden and Persian Gulf ports (a good deal of it destined for Turkey and southern Russia) was recorded during the eighteenth and nineteenth centuries, and it was still at or near the top of the list of imports into Baghdad in the first decade of the twentieth century.[40]

Last sample in the Beaulieu manscript, the earliest detailed description, with samples, of Indian cotton dyeing and painting, compiled in Pondicherry on the Coromandel coast, early 1730s (see also pp.158–9).

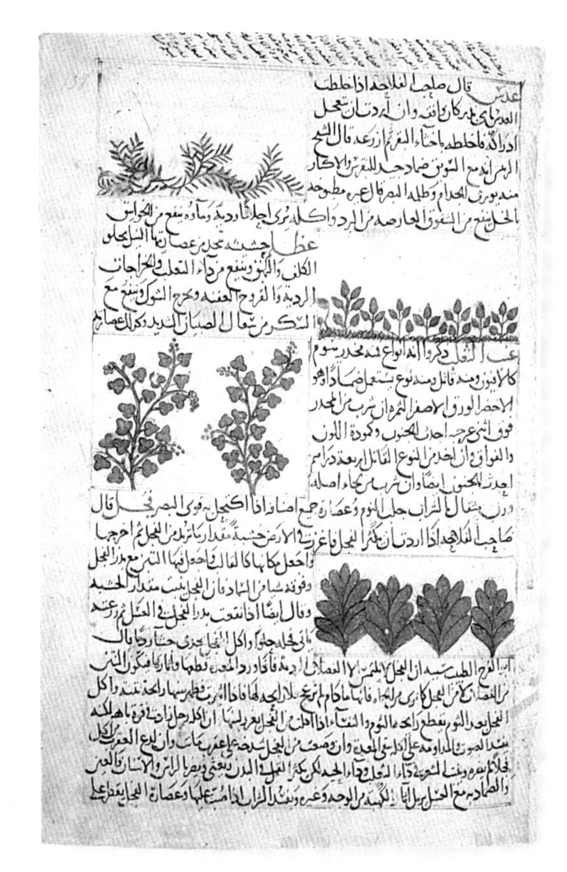

Qazwin manuscript, Persia, sixteenth century, illustrating (top right) the indigo plant.

During the late sixteenth century French, English and Dutch traders had greatly increased their activities in the Levant, cutting out Venetian middlemen as early as the 1560s. Thereafter, with protection as well as concessions granted by the Ottoman authorities, European traders had organized themselves into official Levant companies. The rivalry between the routes they used and the Cape route inevitably intensified.[41] The safety of either route was always unpredictable: at sea piracy and shipwreck, and on land raids on caravans, were a constant threat and caused wild fluctuations in the selling prices of Asian goods. As the seventeenth century advanced, the Cape route became better established to the point where goods brought to Europe that way were cheaper than goods brought on the Levantine route. This would have curious consequences for European trade in the Levant in certain special commodities, in particular indigo.

From their foundation the Levant companies were much concerned with indigo. Correspondence between 'factors' (agents) of these companies in Aleppo and their headquarters in Europe in the final decades of the sixteenth century frequently emphasized the importance of indigo, and its exceptionally high value pound for pound compared with nearly all other 'spices'. It was usually sold by the 'churle' (Italian *zurli*, i.e. cowhide bundle), packed two to a chest. Although dyestuffs, as already noted, were generally lumped together under the generic term 'spices', indigo and sometimes lac (or its Spanish American replacement, cochineal) were often itemized separately in agents' reports and bills of lading.[42] A Venetian consul, for example, in an overview of Syria's trade, records that 'many nations from many divers parts bring hither a vast array of very rich commodities, spices, silk, indigo, cotton, cloth made of wool, silk and gold, and infinite other things'.[43] From the end of the sixteenth century the raising of heavier duties by the Ottomans on the dye trade gave further impetus to the European shipping companies to engage in the long sea route to the East.

In 1599 alarming news reached Aleppo that six Dutch ships had arrived in Holland from the East Indies laden with oriental goods. To Levant companies this sounded 'catastrophic'.[44] However, once they learnt that the Dutch were concentrating on pepper and cloves they breathed a deep sigh of relief, merchants' correspondence declaring that as long as the Dutch did not carry much indigo, nutmeg or silk the Levant trade would be little harmed. One letter tells of the cargo of the *Hector*, as she prepares to sail from the Levant to London at the end of that year:

Resist-printed linen *indienne*, dyed in indigo. French, eighteenth or nineteenth century.

Ther will goe good store of all commodities upon hir but especially indico and silke, all men havinge advise cheefly of those commodities, being fearfull of spices, by reason of the newes of the Flemmings discovery in the East Indeys; as also those four ships arrived in Amsterdam having brought great store of pepper and cloves, none cares to meadell with these commodities.[45]

Pew end of a fuller at work with a two-handled mallet. Above is a weaver's beam with cloth and there are the tools of this valuable trade – shears, knife, weaver's comb, and teasel holder. Spaxton church, Somerset, sixteenth century.

Sources like Robert's great *Merchants Mappe of Commerce* confirm that from the early seventeenth century supplies of indigo were plentiful in the Levant. By 1630 the Cape routes had reduced transport costs to such an extent that Indian indigo could now be brought to Europe by East India companies, re-exported from there to Levantine markets by the Levant companies, and sold or bartered at a lower price than indigo brought overland.[46] In England's case, this saved an undesirable drain on the country's bullion. (The EIC's original aim of obtaining oriental spices by bartering surplus English woollen broadcloth had proved unsuccessful, but at least the ability to re-export such suitable goods as Indian indigo reduced the drain on England's currency.)[47] In the following century the English Levant company was officially prohibited for fifty years from exporting any bullion from England, which put it at a disadvantage *vis-à-vis* its French and Dutch rivals, who had more flexibility. The English government did at least designate certain colonial products, including indigo, 'money goods', which could be exchanged for local currency if necessary.[48]

The French, with their large supplies of West Indian indigo and sugar, were particularly enthusiastic about the re-export trade. From the mid-1680s West Indian indigo was the most 'interesting' of all commodities arriving at the ports of Bordeaux and Marseilles.[49] A hundred years later a French traveller reported that France had the greatest trade to Syria of any European nation. Re-exported indigo was still one of her top commodities traded there, although by then overtaken by cochineal.[50]

It must not be forgotten, with so much talk of trade criss-crossing the Mediterranean and the Middle East, that untold quantities of dyestuffs were being used locally throughout the Ottoman Empire and beyond, to produce everyday and luxury textiles, including tribal rugs and the finest court carpets. Of the general consumption of dyes in Anatolia, for example, one historian of Ottoman archives notes that: 'Among the dyes employed by sixteenth century Anatolian dyers, indigo was most frequently referred to in contemporary documents. It was an expensive raw material.'[51] Natural indigo, and its synthetic successor, continued to be a major import into Ottoman ports into the twentieth century.

BLANC·TEINT·LES·DRAPS
ET·INDIENNE·EN·TOUTES
SORTES·DE·COULEURS
A·JUSTE·PRIX·A·SAIN·BEL

Europe's quest to understand the new dyestuffs

In the second half of the sixteenth century, while Cape trade was in its infancy, there was a general impetus in Europe not only, as we saw in the last chapter, to seek fresh sources of woad but also to improve dyeing skills at home by finding out more about the intriguing new dyestuffs appearing on the markets. Nations competed both for access to, and understanding of, indigo and logwood, which were increasingly being sought as alternatives to woad. There was a dawning realization that indigo was better than woad, but there was a long period of experimentation with dye plants and methods, and of considerable confusion about them.[52]

Attempts were made by the Spanish in the 1550s and the English in the 1580s – both keen to bypass the French product – to acclimatize woad to the New World of the Americas. Yet in the same period chroniclers and traders from both nations were expressing interest in a native indigo plant being used by the American Indians,[53] and even engaging in some trade in 'annele (which is a kind of thing to dye blew withall)',[54] but were uncertain exactly how the dye was used. However, it was not long before indigo's potential was realized. In 1558 the Spanish king expressed keen interest in the prospects for indigo, and soon afterwards the Spanish were introducing new species of indigofera into their American colonies and rapidly developed highly productive manufacturing methods.

England's quest for new ideas are exemplified in Hakluyt's *Principall Navigations*. He stressed that English cloth would not find ready markets abroad nor would problems of unemployment be solved, unless home dyers sharpened up their skills:

For that England hath the best wool, and cloth of the world, and for that the cloths of the Realme have no good vent if good dying be not added: therefore it is much to be wished, that the dying of forren Countreies were seene, to the end that the arte of dying may be brought into the Realme in greatest excellencie; for thereof will followe honour to the Realme, and greate and ample vent of our clothes; and of the vente of clothes, will followe the setting of our poore in worke, in all degrees of labour in clothing and dying.[55]

To this end in 1579 Hakluyt dispatches Morgan Hubblethorne, a master dyer, to Persia on a mission to find out how the Persian dyers achieve such excellent results. He also instructs him to keep his eyes open *en route*: 'You must have great care to have knowledge of the materials of all the Countreies that you shall passe through, that may be used in dying, be they hearbes, weedes, barkes, gummes, earths, or what els soever.' The instructions continue:

In Persia you shall finde carpets of coarse thrommed wooll, the best of the worlde, and excellently coloured: those cities and townes you must repaire to, and you must use meanes to learne all the order of the dying of those thrommes, which are so died, as neither raine, wine, nor yet vineger, can staine: and if you may attaine to that cunning, you shall not neede to feare dying of cloth, for if the colour holde in yarne, and thromme, it will holde much better in clothe ... For that in Persia they have great colouring of silkes, it behoves you to learne that also.

Hubblethorne is also encouraged to seek out Chinese dyers in Persia and in particular to investigate indigo plants and production methods:

You shalle finde Anyle *there, if you can procure the herbe, that it is made of either by seede, or by plant, to carry into England, you may doe well to endevour to enrich your countrey with the same; but withall learne you the making of the* Anyle, *and if you can get the hearbe, you may sende the same drye into England, for possibly it groweth here already.*

The belief that such indigo might grow in England demonstrates the haziness of knowledge about foreign dyestuffs at that time, as does a further instruction to look for logwood in Persia: 'Learne you there to fixe, and make sure the colour to be given by loggewood: so shalle we not neede to buy oade [woad] so deere, to the enriching of our enemies.'

Hakluyt's hopes of finding logwood in Persia and using it as well as indigo to replace woad were false on two counts. Firstly, logwood only grew in America, as the Spanish had recently discovered. Secondly, the blues it produced were not colourfast (although mordanting methods did later improve). It did, however, make a black so valuable in commerce that it was the cause of many conflicts between the Spanish and the English in Central America, and even resulted in the creation of British Honduras (Belize) – but that is another dyestuff story![56]

The woad war

The producers of woad would hardly have supported Hakluyt's efforts on behalf of indigo. While European merchants were battling on several fronts to corner the market in exotic commodities, including the tropical dyestuffs, all those with vested interests in woad fought back on its behalf. The indigo versus woad battle was fierce and prolonged, in some places lasting well into the eighteenth century. Landowners and local merchants, especially on the European continent, were in confrontation with all the companies eagerly expanding their trading interests in Asia and, later, the West Indies. The woad producers' horror at the prospect of losing their livelihoods foreshadowed that of the expatriate indigo planters in India at the very end of the nineteenth century, equally alarmed by the new threat of synthetic indigo. The rearguard tactics of both were doomed.

Protectionist edicts were issued from the end of the sixteenth century throughout Europe in support of home-grown woad, and a deliberate smear campaign gave imported indigo a bad press. Dyers, moreover, were ignorant about exactly how to use the new dye, which needed different handling from woad, including the use of corrosive vitriol or arsenic. It was also confused with logwood, whose 'false indigo' blue was unstable.

In France the landowning plutocracy of Languedoc had grown fat on woad profits and had little difficulty in persuading the government, which gained much tax revenue from woad, to ban the import of indigo in 1598. Eleven years later the king issued a draconian edict which actually sentenced to death anyone found using 'the deceitful and injurious dye called *inde*'.[57] Later in the century the virtues of indigo were inevitably recognized, notably by the French minister Colbert in 1669, but French dyers were not all officially free to use imported indigo as they wished until 1737.

Daniel Gottfried Schrebers,
der Rechte Doctors,
historische, physische und oeconomische Beschreibung
des
Waidtes,
dessen Baues, Bereitung und Gebrauchs zum Färben,
auch Handels mit selbigen überhaupt,
besonders aber in
Thüringen.
Mit Beylagen, und einem Anhange dreyer alter Schriften.

HALLE,
In Commißion der Buchhandlung des Waysenhauses, 17[illegible]

Illustration of a German woad mill in Thuringia from Schreber's famous book on woad, published in 1752.

Other European countries followed suit. In Germany, despite the damage inflicted on its agricultural economy by the Thirty Years' War in the first half of the seventeenth century, the woad industry remained an important source of wealth. From 1577, and for over one hundred years thereafter, various imperial and local prohibitions were announced banning indigo, 'the devil's dye', on the grounds that it was 'pernicious, deceitful, eating and corrosive'.[58] A typical decree issued at Dresden in 1650 gives a flavour of the prevailing strength of feeling in Thuringia:

By the Grace of God ... it is known to each and all of you that our Province of Thuringia has been blessed by the Almighty above all other countries and provinces with the Woad plant ... Cloths and other fabrics of a good quality were dyed [in woad], everyone being satisfied both with their quality and durability. On the other hand there is clear proof that indigo not only readily loses its colour but also corrodes cloths and other fabrics, thus causing serious loss to many worthy persons ... We therefore command you . . . to prohibit under pain of confiscation, the sale of any cloths and other similar articles which are not dyed with Woad, but with other injurious dyes . . .We also publish this express Commination that, if any person shall deal in any such deceptive dyes or similar wares or import the same, we shall severely punish him . . .[59]

At the close of the eighteenth century the magistrates of Nuremberg were still forcing their dyers to swear annually under oath not to use indigo, and, like the French authorities, actually threatened dyers with the death sentence for disobedience. The edict concerned was still on the books at the end of the nineteenth century but had long since been ignored.[60]

In Britain there were fewer vested interests at stake, but from 1532 imported indigo was nevertheless denounced as 'food for the devil', and tropical dyes were subject to prohibitions. In 1581 (just two years after Hakluyt's instructions, quoted above, pp.54–5) an Act was passed which authorized searchers to burn any logwood found in a dye-house, and allowed the use of indigo ('a nele, alias blue Inde') only as an addition to the woad vats to make a bottom dye for blacks. Soon afterwards, though, opponents of indigo declared it to be poisonous, and its use was forbidden by law until 1660. However, twenty years before the ban was lifted dyers were being encouraged to switch to indigo. Parkinson, in his *Theater of Plantes*, fully realizes its potential, although he relies on hearsay to describe tropical indigo, which he calls 'Indico or Indian Woade':

Although Nil *or* Indico *be not in forme like Woade, yet for the rich blew colour sake I think good to make mention of it here, not only to show you what it is, and how made, but to incite some of our nation to be as industrious therein as they have beene with the former Woade, seeing no doubt that it would bee more profitable.*[61]

For the rest of the seventeenth century the superiority of indigo came increasingly to be recognized. European traders clearly bypassed the laws and dyers were keen to switch to indigo long before their governments gave them official sanction to do so. The legislation promoted by the woad lobby proved economically impossible to maintain, although the ban on logwood continued well into the eighteenth century.

But although all protectionist measures ultimately failed, foreign indigo, even when legitimately imported in quantity, did not entirely eclipse woad. The latter played a new role as a fermenting agent in

Engraving of an indigo dyeworks in Egypt made during Napoleon's occupation of the country from 1798. On the floor clay vessels contain ground indigo mixed with water. On the platform a dyer stirs one of the vats beneath a line of dyed cloth.

indigo vats even into the twentieth century, mainly to dye quantities of 'navy blue' worsted for government and service uniforms throughout Europe.

There was even one extraordinary attempt to resuscitate the woad industry in southern France, Austria and Italy. When in 1806 Napoleon set up his 'Continental System', which excluded England's ships from European ports in an attempt to wreck her trade, England retaliated by using her naval power to prevent foreign ships and colonial goods from reaching France. Deprived thereby of imported tropical indigo dyestuff, so necessary above all for dyeing his army's uniforms, Napoleon in desperation offered large rewards to anyone who could find an effective local substitute to replace it. Trials with tropical indigo, grown in Spain, failed,[62] as did experiments to make Prussian Blue (see the Glossary) colourfast on woollen cloth.[63] As the most likely candidate was woad, a decree went out that this should be planted throughout France and in other parts of Europe. Thousands of acres were planted with woad, and elaborate research to try and isolate indigo dyestuff from the leaves (hitherto only composted leaves had been used in the dye vat) was carried out and published in several works.[64] Existing descriptions of indigo production in the colonies and in Egypt, the latter being compiled for the *Description de l'Egypte* commissioned after the French invasion of Egypt in 1798,[65] must have been of great interest. Nevertheless, the project failed, since extracting indigo from woad leaves is difficult. The continental blockade was lifted in 1812 and trading in indigo soon reverted to normal.

Indigo manufacture, French West Indies, late seventeenth century: harvesting in the backgound, steeping tanks to the right and beating apparatus in operation below them. On the left, indigo paste in conical bags is being carried to the drying shed.

European colonies and their indigo plantations

In the seventeenth century demand in Europe for indigo was expanding fast. Growing conditions in the Indies were ideal, as they were for sugar and rice, and the new colonies offered massive scope for exploitation. If a country was found to have a local tradition of making indigo this indicated that conditions would be suitable for commercial plantation cultivation linked to large-scale extraction systems. Contemporary reports on the economic potential in all of the colonies are full of references to indigo. This applies not only to the larger ones – in the Dutch East Indies, the French West Indies, Spanish Central America, British America and India – but even to small ones like the tiny Malagasian island of Nosy Bé (capital, Hell-Ville), hardly a household name today but meriting several pages of its own in a nineteenth-century publication which describes the 'useful plants' of the French colonies.[66]

De l'Inde & Indigo, & de la maniere qu'ils se fabriquent.

Producing indigo is labour-intensive, but the problem was solved by coercion of the local labour force or, so far as the West Indian and American colonies were concerned, by importing African slaves, considered as expendable commodities. This involved indigo in an elaborate 'Dance to the Music of Time' across many nations. Four-cornered trade brought Asian indigo-dyed textiles to Europe for re-export to West Africa as essential barter for slaves. These slaves provided much of the labour on the European indigo plantations in the West Indies and America, and the end product crossed the Atlantic for consumption in Europe.

Without a suffering labour force indigo's story would be very different. For every five acres of indigo four labourers were required. As the minutes of the council of the French colony of Louisiana explained in 1723: 'There are few inhabitants in a position to undertake the culture of indigo for which a sufficient number of negroes is needed, otherwise it must not be thought of ... it is for the gentlemen of the Company to hasten its cultivation by a prompt dispatch of negroes.'[67] This refers to the French 'Company of the Indies', to whose directors a plea was addressed the same year for more slaves to be sent: 'Everybody, Gentlemen, is asking for negroes in order to be able to have work done on preparing land to plant indigo for next year.'[68] One French source on indigo making decries the 'malice' of the 'negro' in the most patronizing terms but admits that 'we calculate our revenues only by the number of slaves of both sexes who are employed in our manufactures'.[69] Later, when prices were high, indigo dyestuff could be exchanged for slaves; it was said that a planter in South Carolina could fill his bags with indigo and ride to Charleston to buy a slave with the contents, 'exchanging indigo pound for pound of negro weighed naked'.[70]

Workers, whether slaves or local peasantry, paid a high price, often with their lives, to satisfy the European appetite for blue. For this reason indigo's history is tainted by the associated evil of human exploitation; and this touches us even today as we innocently admire the end-products of wealth amassed from former staple colonial crops such as indigo. How many people, for example, know that the USA world golf championship in Augusta, Georgia, is staged on a former indigo and cotton plantation?

West African slavery and trade

In West Africa indigo features in the interlocking pattern of religion, trade and slavery. The expansion of West Africa's indigenous textile industry and trade had been stimulated by the spread of Islam, which brought with it a desire for clothing. In particular the major Islamic centres of Timbuktu and Djenné both produced cotton cloth and for centuries traded extensively

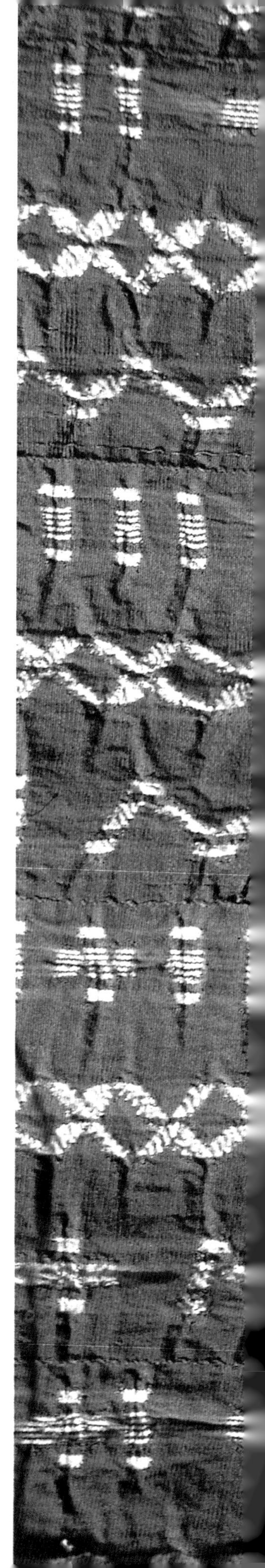

across the Sahara with the Mahgreb and Europe. In the Middle Ages much exchange of cloth – including dyed cloth and dyestuffs – revolved around trade in salt, ivory and gold.[71]

From the middle of the fifteenth century the Muslim monopoly on West African trade was broken by sea-faring European traders who focused their attention on coastal trading posts. Textiles comprised the dominant item in a complex system of barter until the abolition of the slave trade in the nineteenth century; and many were indigo-dyed to meet local demand.[72] The Portuguese and Spanish were quick off the mark in the race to trade in Africa, but it was the Portuguese who were to dominate in much of the sixteenth century. The main objective was to obtain elsewhere goods suitable for barter in the interior of West Africa. Not only did they acquire Indian textiles, but they also produced and dyed cotton cloth in their nearby Cape Verde island colonies, using imported African labour and skills to cultivate and process large quantities of indigo and cotton.[73] Portuguese agents also procured extra indigo dye in mainland West Africa, some from the interior of Portuguese Guinea. When Portugal was subordinated to Spain from 1580 to 1640, the Spanish capitalized on Portuguese mercantile expertise to supply slaves to service her by now enormous empire in America and the Indies. Spain was also trying vainly to force out her European rivals in Africa.[74]

Snapping at the heels of the Portuguese, French and British traders had arrived on the coast of West Africa in the early sixteenth century. At the end of the century they were followed by the Dutch. However, despite intense rivalry, no one managed to establish a monopoly. They all acquired African cloth in vast quantities, often in exchange for Indian and European cloth, in coastal trading centres of the 'Guinea coast', especially the Senegambia river ports and the Niger delta in the Gulf of Guinea.[75] Everyone sought slaves above all, but also such goods as salt, slaves' clothing and boat awnings, as well as ivory and gold. After the 1630s the re-export of Asian goods from Europe for marketing in West Africa was handled by Europe's various African shipping companies. Any surplus cloth remaining after procuring slaves was taken on to the West Indies to be used as slave clothing or exchanged for plantation produce. Particularly popular in Africa were the so-called 'Guinea' or 'long' cloths, most being plain or patterned indigo calicoes from India's Coromandel coast.

Indigo-dyed textile with a circular design still known as 'slave shackles'. Acquired in 1997 in the river port of Bakel, Senegal, where handwoven cotton strip cloth is still dyed in natural indigo.

A French traveller in Guinea in 1818 commented that the inhabitants of Fouta Djallon 'wear cloth of blue guinee stuff, which is a sign of great luxury among these Africans',[76] and the Englishman Mungo Park compared such cloth with the local version.[77] Later in the century in Senegal the French were still importing from French Pondicherry indigo-dyed *guinées* (also called *guinées bleues* or bafts). These were used as currency,[78] although their value fluctuated wildly and their dimensions were not standardized until the 1870s.[79] (Countries of the western Sahara adopted the term *guinée* to describe plain indigo-dyed cloth from whatever source.) By the second half of the eighteenth century manufacturers in England and France were also successfully imitating the popular Indian cloths; their cotton checks called 'Guinea cloths' enjoyed great success in Africa. In fact, the freed slave Olaudah Equiano concluded his extraordinary autobiography of 1794 with a plea for the abolition of slavery on the grounds that it would enormously benefit such British manufacturing interests. He wrote:

If the blacks were permitted to remain in their own country, they would double themselves every fifteen years. In proportion to such an increase will be the demand for manufactures. Cotton and indigo grow spontaneously in most parts of Africa; a consideration this of no small consequence to the manufacturing towns of Great Britain – the clothing etc. of a continent ten thousand miles in circumference, and immensely rich in production of every denomination in return for manufactures.[80]

In the 1680s the British in Sierra Leone explored the possibilities of producing their own indigo there.[81] Much later the French, when their slave trade was under attack in the early nineteenth century by the British, sought new sources of revenue. They initiated projects to exploit indigenous African crops, especially cotton and indigo, which had brought such profits to West Indian planters. They had also since 1787 been successfully exporting excellent indigo from their settlement on the island of Gorée, just south of Cape Verde.[82] One example of an indigo initiative took place in the Senegalese riverside town of Richard Tol ('Tol' meaning 'garden'). Here the governor, Baron Roger, in collaboration with a horticulturist called Richard and several French chemists, obsessively experimented with the cultivation of indigo, with great success for a time, as conditions were ideal.[83] The French explorer René Caillié remarked at that time on the quantities of indigo and cotton growing beside the Senegal river as he headed upstream on his journey to Timbuktu.[84] Following the same route today, the crumbling French forts, mansions, quays and waterside warehouses of sleepy towns lining the Senegal river are testimony to this harsh but lively episode of colonial economic history.

Central and South America

Although it was not until the middle of the seventeenth century that the East India Companies really began facing serious competition from New World indigo, the Spanish had begun to exploit the dyestuff there almost a century before. Although the export of minerals was to be of enormous importance, trade in dyestuffs – cochineal, indigo and various dyewoods – was to bring the Spanish, as it did the French, British and other European settlers in the West Indies and North America, much wealth.

Once established in America, the Spanish had, as already mentioned, noted with interest that the Aztecs and Mayas were using indigo dye from native indigoferas.[85] Indeed, they traded small quantities of Indian-made indigo between their American territories;[86] but keen to bypass Portuguese supplies, they introduced Asian *Indigofera tinctoria* to Central America in the second half of the sixteenth century.[87] Indigo was first produced on a commercial scale in Guatemala (conquered by the Spanish in 1524) and in regions grouped under its captaincy-general: El Salvador, Nicaragua, Honduras and Chiapas. It also became a profitable export from Mexico and parts of South America – Caracas (Venezuela), Ecuador and Peru – and for a time from Spanish territories in the West Indies. The Portuguese later had an indigo boom in Brazil (see below).[88]

The Spanish conquistadors, disappointed that Guatemala did not possess minerals, turned to agriculture as a source of wealth. Indigo, sugar-cane and tobacco proved to be ideal crops when the cacao industry was declining towards the end of the sixteenth century. Indian communal lands in fertile agricultural areas, especially along the Pacific coastal plain (now El Salvador and Nicaragua) but also inland, were appropriated by the colonialists.[89] Hundreds of well-organized indigo dye works were established, and by 1600 indigo – known as *xiquilite* or *piedra de añil* – already formed the mainstay of the Central American economy.

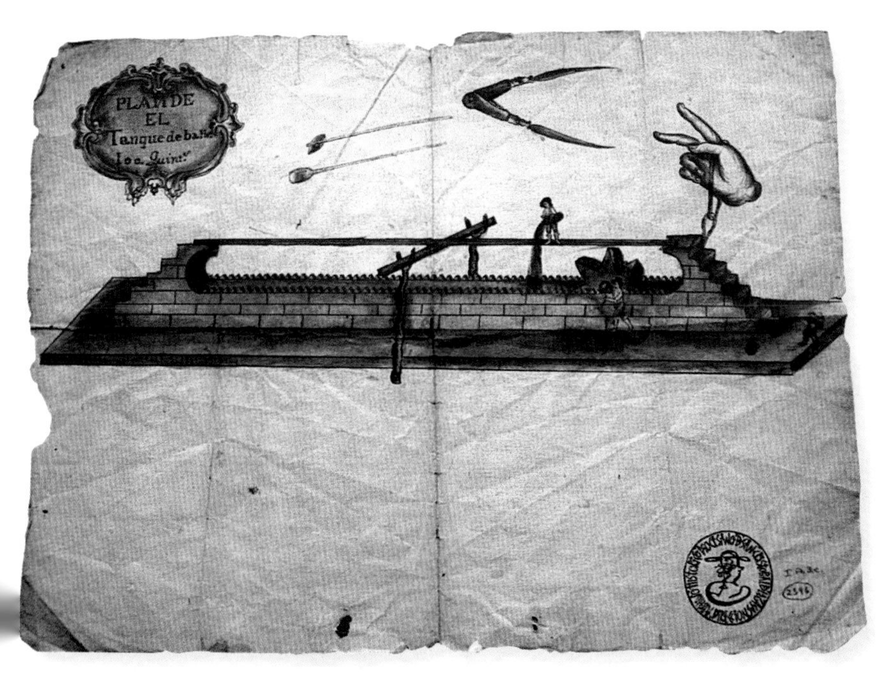

Tank for aerating indigo liquid, using a hand-driven paddle, during colonial manufacture of indigo in Mexico. Illustration in an eighteenth-century manuscript.

The boom time for Spanish colonial indigo was the eighteenth century, when in peak years well over a million pounds of dyestuff were being exported to Spain itself and more to Peru and Mexico. Like the French West Indian product, much of this Spanish indigo was destined for re-export (despite incurring an extra 20 per cent duty), England and Holland being significant consumers.[90] 'Guatemalan' indigo was considered the best quality. As on the East India trade routes, merchants faced many hazards: piracy, sea battles with European rivals, or storms could cause the loss of a year's export of indigo. Cargoes recovered from Spanish galleons and 'West Indiamen' sunk *en route* to Europe from America and the West Indies have included wooden chests, barrels and earthenware jars still filled with indigo. Amongst the cargo from a Portuguese ship wrecked in 1621, for example, were found treasure, hides, *lignum vitae*, and twenty large mahogany boxes of indigo still intact.[91] It was commonplace for cargoes to include over fifty huge barrels of indigo.[92] Such losses caused price fluctuations which, coupled with high taxes, made indigo a volatile commodity to deal in. From the middle of the eighteenth century, however, purchasing prices of the three main grades of indigo (*corte*, *sobresaliente* and, the best, *flor*) were in theory fixed by arbitration at an annual fair attended by Guatemalan and Spanish merchants and growers' representatives. An Indigo Growers' Society was formed in 1782 (abolished in 1826) in an attempt to regulate the whole indigo industry.[93]

The growers themselves had two main headaches, in addition to dread of earthquakes and unpredictable seasonal rain. One was contending with insect damage, particularly that caused by caterpillars and plagues of locusts – this gave rise to the expression 'planters of indigo go to bed rich, and rise in the morning totally ruined'.[94] The other constant problem was maintaining enough workers for such a seasonally labour-intensive and unpopular occupation. The smell of fermenting indigo attracted clouds of flies, making both workers and their animals ill, sometimes fatally so, although increasing mechanization of the beating stage did at least improve conditions. From 1581 the Spanish authorities sought to protect the indigenous Indian population (as distinct from imported slaves) with a series of measures and appointed official inspectors to enforce them; but such were the profits from indigo that corruption was widespread. In 1630 some growers even tried to bribe the king personally to suspend inspections.[95] In practice Indians were used to augment African slave labour, which was expensive and in short supply. Codes for dealing with labour problems were always hard to enforce and the crown restriction on the use of Indian labour was lifted in 1738.

Revival of indigo manufacture in Chalatenango, northern El Salvador, using tanks constructed during Spanish colonial times. Packing down the cut indigo plants to keep them submerged, 1996.

Early in the nineteenth century indigo production was declining in Guatemala, due to heavy taxation and increased outside competition, and the Spanish government encouraged diversification into other crops, notably coffee. After independence in 1822 Guatemala's conflict with neighbouring El Salvador reduced her indigo production still further, while cochineal production, on the other hand, expanded. Nevertheless, although coffee became the main item of Central American trade, Honduras, Nicaragua and El Salvador, all continued to export indigo in commercial quantities until synthetic indigo undermined the market.[96]

In the nineteenth century indigo production in El Salvador was almost as important to the economy as coffee is today, and well into the twentieth century the country was supplying her neighbours with natural indigo. The final export of ten tons to Peru is said to have taken place in 1972, though attempts are being made to revive the industry.[97]

Mexico was also well known for indigo production both under Spanish occupation and into the late twentieth century. The most noted regions were Michoacàn and Chiapas,[98] and a little natural indigo was still being grown in the 1990s in Oaxaca. Recent projects to breathe new life into Mexico's natural indigo industry have been undertaken around Juchitan and Niltepec on the Isthmus of Tehuantepec.[99] Purists in Europe and elsewhere who insist on using only natural indigo have recently relied on Mexico as their source, the quality and distribution of indigo from India being unpredictable.

Although not as noted for indigo production, colonial Brazil at one stage had a significant indigo phase of its own.[100] In the early seventeenth century Portugal was still concentrating on woad production in the Azores, but after the 1640s various rather unsuccessful attempts were made to cultivate indigo in commercial quantities in Brazil. However, from the 1750s Portugal was disturbed both by declining state revenues from gold, minerals and sugar, and by having to rely on her adversary Spain for indigo supplies. Accordingly, steps were taken to boost Brazilian production. By the 1770s keen growers in southern Brazil were neglecting their food crops in favour of indigo; and by the 1780s 406 indigo factories had been established in the Rio de Janeiro region alone and there were food shortages in the city. A decade later indigo exports reached a high of 264,768 pounds, enough to fulfil Portugal's requirements and re-export over a third to other European nations. From then on, however, owing to political developments production fell. As Alden puts it: 'It is ironic that Portugal, which had been the first to bring *Indigofera tinctoria* to Europe by sea, should find her efforts to make Brazil into another India thwarted by the rebirth of the Indian indigo industry!'[101]

The West Indies

From the middle of the seventeenth century the French and the British also established in the West Indies extensive plantations for indigo, as well as for sugar, coffee and cotton. All relied increasingly on imported slave labour. As indigo required less capital outlay than sugar, the industry expanded rapidly, especially in the islands of the Antilles.[102] By 1672, just seventeen years after taking over Jamaica from the Spanish, the British had already established sixty indigo plantations on the island (and others on Barbados and Montserrat),[103] but high export duties soon caused planters there to abandon indigo in favour of sugar and coffee. Britain turned instead to her colonies in North America, notably South Carolina, for supplies of middle grade indigo, relying for top quality largely on Spain's Guatemala and France's Saint Domingue (ceded to her by Spain in 1697).[104]

In that important colony (now Haiti), and also in Guadeloupe and Martinique,[105] French production of indigo escalated in the eighteenth century, apart from interruptions caused by the War of Jenkins Ear in 1739–48, and the Seven Years' War which began in 1756. Records show that Saint Domingue had no fewer than 3150 indigo plantations by the 1780s,[106] producing hundreds of tons of dyestuff annually. Indigo was among the key commodities arriving in the ports of Bordeaux and Marseilles, destined both for the expanding textile industries of France and for re-export.[107] The French government appointed strings of scientists to explore ways of improving the processing of indigo; in particular, they invented the most elaborate mechanisms for aerating indigo liquid.[108] Indigo made by traditional methods in the East could not compete with indigo produced so effectively. However, the mulatto/negro rebellions in Saint Domingue in the 1790s and the effects of the French Revolution and the Napoleonic wars virtually put paid to the French indigo manufacturing industry.

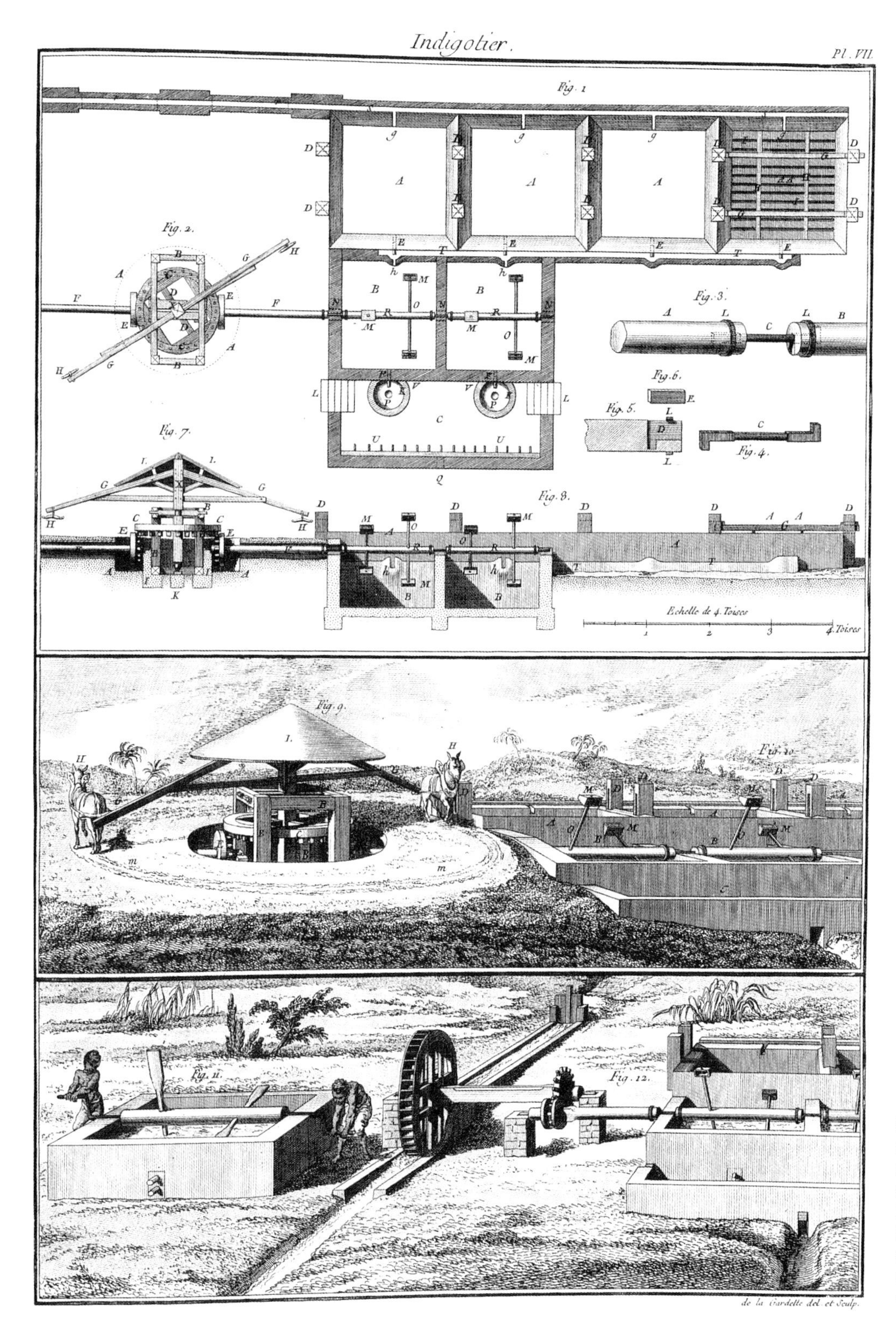

Eighteenth-century French engraving showing various methods devised to turn the paddles for aeration of indigo liquid during dye manufacture in the West Indies. Top and centre show a horse-powered system, bottom right a water wheel, and bottom left manpower.

The southern United States

Dutch settlers are said to have tried to produce indigo from wild indigo plants near present-day New York in the 1650s,[109] and there are references to indigo growing generally soon after this in the southern United States, although there is much confusion between cultivated species and those growing wild. At this period, when settlers from different European nations were experimenting with a range of plantation products, indigo's name frequently turns up along with other potentially viable cash crops such as ginger, sugar, wine grapes, rice and cotton. However, the indigo industry really got going in the 1730s, again thanks to the increased availability of slaves. It combined well with rice as slaves could be fully employed during the slack season for indigo. For the rest of the century indigo was a major staple crop in the region, the biggest producer of all being South Carolina. Although plenty of indigo was exported to Europe, a good deal was consumed by the home market. A full account of indigo's tentative beginnings, its starring role and subsequent decline into obscurity in the nineteenth century can be found in Gray's *History of Agriculture in the Southern United States to 1860*.[110]

From the 1720s the French government actively encouraged Jesuit settlers in Louisiana to produce indigo by sending over seed from the West Indies and importing the product.[111] Advisers from West Indian plantations helped many private plantation owners to build indigo factories in the lower Mississippi valley around New Orleans, and these were soon producing around 70,000 pounds of dye per annum. Conditions were ideal provided seasonal flooding could be overcome. In the second half of the century the quality of the product had improved, and over 240,000 pounds of dye were being produced. The three main types were sold as *flotant* or *flor*, *gorge de pigeon* (an apt description for the violety colour), or *cuivre* (called in the English colonies 'Fine Flora', 'Fine Purple' and 'Fine Copper'). When Louisiana fell under Spanish control after the Seven Years' War indigo production continued to boom, many settlers being enticed to the region by the prospect of indigo fortunes. Up to 550,000 pounds were sold annually in the 1790s. No wonder many rich planters left their descendants wealthy legacies based on indigo and slaves.[112] Until the very end of the century indigo remained a valuable product, despite marketing problems aggravated by competition from Guatemala, exhaustion of the soil, the increased price of slaves, depressions in the 1790s due mainly to insect damage and blight, and all the upheavals caused by the American Revolution. Thereafter indigo never picked up again, largely giving way in the early years of the nineteenth century to cotton, sugar and tobacco.

Profits were reaped from indigo exports in the other southern states in the eighteenth century. Georgia, Virginia, Alabama and the Floridas were all producers, as well as areas further north.[113] In both Spanish and British Florida in particular indigo was a major staple, becoming for Britain during their occupation the single most important commercial product, as it also was in Georgia which exported over 50,000 pounds in peak years.[114] In colonies under British control planters succeeded in producing reasonably good quality indigo, but it never matched that of Guatemala, where yields were higher, or that of the French West Indies, where far more slaves were available and more cuttings were possible per year. Nevertheless, optimum yields in the southern states produced 80 pounds of indigo per acre.

In South Carolina rice was the main crop until the 1740s when the determination of a young woman named Eliza Lucas Pinckney (b. 1722) put indigo ahead of other crops.[115] At the age of eighteen Eliza, who had moved to South Carolina from the West Indies, decided to produce indigo on the family estate. She accordingly asked her father, then Governor of Antigua, to send more seed and an overseer to set up the manufacturing process. Despite initial problems, Eliza (now married to Mr Pinckney) harvested a good crop of seed in 1744 and distributed it among other planters. From these beginnings the industry expanded rapidly, as export figures show.

The planters of South Carolina were helped by lucky timing. Jamaica had turned to sugar, prices of rice had fallen and the War of Jenkins Ear had curbed indigo supplies from the French West Indies in the 1740s. From 1748 production soared in Britain's American colonies, stimulated by the British government's payment of bounty (a bonus system also used by the French in Louisiana) to planters who produced good quality indigo. Figures for Charleston, for example, had risen to a peak of over a million pounds of indigo – 35 per cent of South Carolina's total exports – on the eve of the Revolution in 1775.[116] Thereafter, the withdrawal of bounty and the general disruption of trade caused indigo production to decline everywhere.

Material benefits of the mid-eighteenth century boom in indigo production and trade are still apparent today. For example, indigo was largely responsible for the development of Georgetown. In 1755 the Winyah Indigo Society (which still exists) was formed there by men of influence specifically to promote indigo and the social benefits that would result. In the words of one recent historian: 'The more information on indigo, the more profit; the more profit, the more money for education. Thus, the crop would be the basis for a higher culture.'[117]

On a less cultural level insatiable demand for plantation labour led to high prices for slaves. In 1754 the Governor of South Carolina reported that 'negroes are sold at higher prices here than in any part of the King's dominions . . . a proof that this province is in a flourishing condition . . . I presume 'tis indigo that puts all in such high spirits.'[118] As well as requiring slaves for its production, indigo was also one of the commodities used to buy them. In the 1790s, although the trading of slaves between the Caribbean islands was now illegal, traders would take indigo or rice from the coast of Georgia to exchange for slaves in the Windward Islands. These slaves would then be sold on at a profit in Cuba.[119]

During the American Revolution (1775–83) Britain's supply routes were disrupted. Soon after, the loss of her American colonies and the drying up of French supplies of indigo obliged the British government to press for a return to India as a source. This in turn led to depressed prices of indigo in America while prices for rice and cotton were rising. Until the Civil War in the 1860s, bringing with it the abolition of slavery in the South, small amounts of indigo nevertheless continued to be produced commercially in South Carolina and Georgia. Thereafter the industry virtually ceased in America.

Back to India – British indigo production in the nineteenth century

In India the story of indigo's extraordinary rise and fall is every bit as turbulent and colourful as elsewhere. The treatment of the labour force caused enough discontent to have direct political consequences on British rule in India. This crucial final chapter in natural indigo's economic fortunes is well documented in such works as Watt's *Dictionary of the Economic Products of India* and Kling's *The Blue Mutiny*. There is a wealth of material in London's India Record office, and many memoirs of planters and civil servants provide personal insights into the period.[120]

In the nineteenth century Bengal, and later Bihar to its north, became the world's main source of indigo, which was by now in great demand to supply the textile industries of the Industrial Revolution and to dye many European armed service uniforms. Throughout the century natural indigo was far more valuable than any other dyestuff, even synthetic alizarin.[121] Bengal's indigo production far outweighed that of the rest of the world, substantial as this was.

When Clive conquered the vast territories of Bengal in 1751 he couldn't have foreseen how significant this would be for the future of indigo. Before long, however, the region was providing tailor-made land to help replace American and West Indian supplies. Between 1779 and 1802 the East India Company, which was setting its sights on controlling indigo production in

Watercolour by William Simpson of an indigo factory in Bengal, 1863. Buffalo carts transport freshly cut indigo plants to sets of tanks for steeping. Men aerate the water by hand in the nearest tank and behind is the factory where indigo paste was boiled. Beside the factory a water wheel raises water from the river to fill the tanks.

north-west India too,[122] actively promoted Bengal's indigo industry by providing heavy subsidies and underwriting initial big losses.[123] In 1788 the Company brought European indigo planters over from the West Indies to establish efficient factories,[124] and Bengali indigo soon equalled in quality the finest West Indian product. So successful was the operation that the Company's Asian indigo imports into London jumped from 25,000 pounds in 1782 (when most indigo came from America and the West Indies) to 4,368,000 pounds in 1795, the bulk of it from Bengal. After withdrawing from direct control of the factories in 1802, the Company now paid money advances to private planters but still channelled the lion's share of trade through its 'agency houses'. As everyone wanted to jump on the bandwagon, many Company officers chose to specialize in indigo trade or resigned to become planters. By 1815 Bengal alone exported a staggering 7,650,000 pounds – over 3,500 tons – of indigo (valued at around six shillings a pound).[125]

Bengal's indigo industry was, however, dogged by a conflict of interests, owing to the unsatisfactory way the industry operated there. As European planters were forbidden to lease or buy land adjoining their factories, they had to advance money to reluctant local peasants (*ryots*) through corrupt local middlemen – supervisors, money-lenders and landlords (*zamindars*). Planters were also handicapped by unscrupulous European rivals who sometimes appropriated indigo already contracted out to another planter.[126] The system was deeply unpopular with the peasants as they rarely reaped any benefit from growing indigo, which reduced their rice cultivation, and they were kept in permanent debt.[127] In theory they were free, but in practice they were locked into a system akin to slavery; indeed, one enlightened governor compared their situation with that of Carolina slaves.[128] Peasants could lodge petitions against British planters, but local courts seldom upheld them until later in the century when new laws began to be enforced.[129] Planters used the traditional Indian method of coercion to deal with peasant opposition, i.e. employing local thugs armed with stout sticks or whips to do their dirty work.[130] One source compared planters' behaviour to 'the worst excesses of the most turbulent Irish county'.[131] In 1829 the then Governor-General of Bengal recommended that planters be allowed to take long leases on land in their own names, arguing that it would benefit both planter and peasant. At home directors of the Company strongly opposed such a move, fearing that uncouth planters, virtually the only British non-officials in India apart from missionaries, would become too dominant and damage Britain's reputation in India. Some legislation was passed in the planters' favour, but the majority abused their new powers as landlords; the reimposition of restrictions was too late to prevent the peasant mutiny of 1859.[132]

Before this, when things were going well, fortunes were made from indigo, but the market was volatile owing to speculation and over-production.[133] As Rhind commented: 'There is no article of commerce which fluctuates more in its price, and is of greater variety of quality, than indigo.'[134] Many British families today have ancestors who in the not so distant past either made or lost their fortunes on Indian indigo plantations. The most prosperous years of all fell between 1834 and 1847, when the consumption of Indian indigo in Britain and America doubled, and indigo accounted for almost half the value of all goods exported from Calcutta. Average exports from Bengal in these peak years, when three or four million people were employed in its manufacture, provided four-fifths of total world supplies. A contemporary source sets out some interesting figures for comparison:

The average supply of indigo at present may be estimated as follows: – Bengal provinces, 34,500 *chests, or about* 9,000,000 *lbs.; other countries, including Madras and Guatimala,* 8,500 *chests. Of this there are consumed in the United Kingdom* 11,500 *chests; France,* 8,000 *chests, Germany and the rest of Europe,* 13,500 *chests; Persia* 3,500 *chests; India* 2,500 *chests; United States* 2,000 *chests; other countries* 2,000 *chests; total,* 43,000 *chests, or upwards of* 11,000,000 *lbs.*[135]

After 1847 various factors, political, economic and practical, caused a sharp decline, although many speculators still had high hopes of gaining fortunes from indigo manufacture and trade.

ii APPENDIX, No. I.

(Enclosure.) .. ..

STATEMENT of the Number of LICENCES to proceed to and reside in
tions refused by the Court; and of the Number of the latter granted by

	1814.	1815.	1816.	1817.
Advocates and Attorneys at Law, &c., ...	...	2	...	2
Agents, ...	...	2	1	..
Apothecaries, ...	...	..	...	...
Auctioneer, ...	...	...	...	..
Bakers, ...	...	..	...	1
Bishop's College at Calcutta, ..	..	...	...	...
Broker, ...	...	..	...	..
Cabinet Maker, ...	...	..	...	...
Carver and Gilder, ...	...	...	...	...
Chemist and Druggist, ...	...	...	...	...
Clerks, &c., Mercantile, ..	...	...	1	...
Coachmakers, ...	...	..	...	1
Confectioner, ...	...	...	...	...
Cooper, ...	...	..	...	...
Dentist, ...	...	...	...	...
Distiller, ...	...	..	..	...
European Servants, ...	...	...	...	...
Gardener, ...	...	...	...	1
Grooms, ...	...	1	...	...
Gunmaker, ...	...	...	...	...
Horticulturist, ...	..	...	..	...
Hotel Keepers, ...	...	...	...	...
Indigo Planters, ...	1	...	..	5
Jewellers, ...	...	...	...	...
Leather Manufacturer, ...	...	...	1	...
Linguist and Schoolmaster, ..	...	...	...	...
Lithographic Printer, ...	...	...	...	...
Midwife, ...	...	..	...	...
Milliner, ...	...	...	...	...
Miniature Painters, ...	...	...	...	...
Miscellaneous, ...	...	3	3	7
Mission College at Calcutta, ...	...	..	...	...
Missionaries, ...	4	7	8	9
Music Master, ...	...	...	...	..
Musical Instrument Maker, ..	...	.	...	..
Organist, ...	...	...	...	1
Orphan Institution in Bengal, ...	..	..	...	...
Partners and Assistants in Mercantile Establishments, ...	...	3	6	14
Persons allowed to proceed to their Friends and Relatives, ...	1	4	6	8
Persons permitted to reside, ..	..	...	...	2
Pianoforte Maker, ...	...	...	...	...
Portrait Painter, ...	..	..	1	...
Private Affairs, ...	3	10	10	9
Roman Catholic Bishop, ...	..	...	...	..
Rope Manufacturer, ...	...	1	...	..
Saddler, ...	...	...	...	..
Sailmaker, ...	...	1	...	..
Schoolmasters, ...	...	...	...	1
Shipbuilders, ...	...	...	...	1

Table of those of all professions applying for licences to reside in India between 1814 and 1831. Traders head the list, followed by missionaries, those with private affairs and then indigo planters.

APPENDIX, No. I. iii

. .. . (Enclosure.)

each Year, from 1814 to 1831, inclusive; of the Number of Applica-
ia; with a General Classification as to Trades, &c., of the whole.

1824.	1825.	1826.	1827.	1828.	1829.	1830.	1831.	TOTAL.
10	6	10	6	2	4	2	3	78
5	4	3	6	6	4	1	1	59
1	..	1	..	..	..	..	..	2
..	..	..	..	..	1	..	..	1
...	..	..	..	1	..	..	..	2
..	..	..	..	..	...	..	..	1
..	..	1	..	..	..	..	..	1
..	..	1	..	..	..	..	..	1
..	..	..	..	..	..	..	..	1
...	1	..	2	..	..	..	..	3
3	2	3	1	11	5	2	3	41
..	..	..	1	1	..	2	1	16
..	..	..	..	..	..	..	..	1
..	..	..	..	..	..	..	..	1
..	..	..	1	..	..	..	1	2
..	1	..	..	..	..	..	..	1
3	1	1	1	1	4	5	7	35
..	..	..	..	..	..	..	..	1
..	..	1	..	..	..	..	..	8
..	..	..	1	..	..	..	..	1
..	..	..	..	..	..	..	..	1
..	..	1	1	..	..	..	..	2
4	12	13	11	11	15	6	4	106
..	..	1	..	1	1	..	..	7
..	..	..	..	..	..	..	..	1
..	..	..	..	..	..	..	..	1
..	..	..	1	..	..	..	..	1
..	..	..	1	..	..	..	..	1
..	..	..	..	..	..	1	..	1
..	..	..	..	..	..	..	..	2
11	6	8	11	12	13	10	11	152
..	..	..	..	..	..	..	..	2
7	13	15	14	7	15	14	11	193
..	1	..	2	..	..	..	1	6
..	..	..	..	..	1	..	..	2
..	..	..	..	..	..	..	..	1
..	..	..	...	..	..	..	..	1
14	17	13	14	19	13	20	36	250
5	2	..	..	1	..	3	2	55
2	1	3	2	8	..	..	4	30
..	..	..	..	..	..	..	..	1
..	1	..	1	..	..	..	..	5
4	4	1	7	5	7	6	5	116
..	..	..	..	..	..	..	..	1
..	..	..	..	1	..	..	..	3
..	..	1	..	..	..	..	..	1
..	..	..	..	..	..	..	..	1
1	..	..	..	1	1	1	..	13
1	..	..	1	..	..	..	..	5

Five years of bad weather, which reduced yields so that the peasant was unable even to recoup his advance on indigo, provided in 1859 a catalyst for disastrous events. Tension had been mounting between, on the one side, planters, colluding magistrates and indigo traders and, on the other, those who championed the oppressed peasants – fair-minded British civil servants (whom other Europeans accused of being 'pro-native') and missionaries, and Indian intelligentsia and landowners. In Calcutta there were three recognized interest groups – the Indigo Planters' Association (formed in 1851), the British Indian Association and the missionary societies. Recent reforms generally unfavourable to the peasants had heightened their rebellious mood by the time Sir John Peter Grant, Bengal's new liberal Lieutenant-Governor (government of India having passed from the EIC to the British Crown following the Indian Mutiny the previous year), attempted in 1859 to establish justice for all. Coming at the end of a volatile decade of racial disharmony in northern India, and with an unpopular new Rent Act in force,[136] Grant's sympathetic attitude emboldened peasants to stand up to the planters. Resentment boiled over in the autumn of 1859, with huge demonstrations, rioting and acts of violence in the indigo districts of Lower Bengal.[137] The situation grew so critical that the Viceroy of India, Lord Canning, wrote: 'For about a week it caused me more anxiety than I had since the days of Delhi ... I felt that a shot fired in anger or fear by one foolish planter might put up every factory in flames.'[138]

In 1860 the worried government set up an indigo commission which produced a report critical of the indigo planters but nevertheless also blamed the flawed system and the frequent failure of peasants to honour their contracts.[139] Wide-scale reforms designed to dispense justice in the countryside were introduced, but disputes continued to plague the industry.

Missionary bodies had always been particularly critical of indigo planters. In 1853 one missionary complained to his Calcutta headquarters: 'Surely it is time that the Indian Government put a stop to such inhuman and cruel proceedings. Every indigo factory deserves to be closed, yea, utterly abolished.'[140] From then on Calcutta missionaries, arguing that indigo planters' behaviour damaged the spread of Christianity in Bengal, became increasingly political, influencing opinion in India and at home, where there were endless acrimonious debates in parliament between the indigo lobby and those seeking reforms in India. The sharpest thorn in the planters' flesh was the outspoken Protestant James Long,[141] who had the ear of the Lieutenant-Governor. Thanks to Long the indigo issue became headline news in London in 1861. This was due to the rumpus caused when Long decided to champion a socio-political Bengali play called

Unfinished sepia watercolour of queues of boats laden with indigo and opium leaving Patna on the Ganges, by Col. G. F. White, 1842.

Nil Durpan ('The Indigo-Planting Mirror'),[142] which satirized the bad behaviour of the planters and their wives, while giving voice to the new Western-educated Indian rural middle class which was sympathetic to the lot of the peasantry. For the planters injury was added to insult by the active collusion of the government of Bengal in the play's distribution. Long was in fact indicted for disseminating a 'libellous' work.[143] His high profile trial which resulted in his temporary imprisonment, and his unrepentant counter-attack in a publication called 'Strike, but Hear',[144] aroused a sympathetic press in both Calcutta and Britain, destroying any remaining public sympathy at home for the indigo lobby. The whole episode, particularly the behaviour of Long and Grant, had a lasting influence on government policy and on the support of the Bengali intelligentsia for the peasantry.

Meanwhile, the beleaguered indigo planters, who saw themselves in a quite different light, published in London a spirited defence of their industry. They claimed that a smear campaign had been launched against them by a government bent on promoting opium. Opium, unlike indigo, was indeed a British government monopoly of enormous value in the China trade. Government (formerly Company) officials and indigo planters were natural enemies; the latter were generally considered rogues, but when indigo reigned supreme planters' activities had been tolerated. However, conditions of trade had altered by the 1850s. British manufactured goods, notably Manchester cottons, had severely affected India's trade, for such items were now imported by India. The staples of India's exports were now all agricultural products: opium, indigo, raw cotton and sugar. Indigo no longer headed the list. Opium now formed 36 per cent of the total and raw cotton 20 per cent, while indigo was left in third position with 11 per cent.[145] The planters' appeal, aimed at 'The British Government, Parliament and People for Protection against the Lieut-Governor of Bengal',[146] accused Grant of inciting indigo labourers to rebel while hush-ing up peasant discontent over the growing of opium. They contrasted their own admirable work with that of the opium industry. The document asserts that Grant, 'for no other object than to satisfy the instincts of a traditional hatred', was trying to destroy an annual trade of £2,000,000 and put a million Indian labourers out of work. Thus, continues the appeal, 'the whole class of planters, who are now reclaiming the wilderness, civilizing the people, curing the sick, etc . . . is to be ruined and driven forth.' In good environmental terms they claimed that 'indigo has cleared the jungle and turned the wilderness into corn-fields, and the lair of the wild-beast into villages; while opium has only covered rich arable land with poppies.'

By the time of Grant's retirement in 1862, protective laws recognized peasants' rights, but old wounds festered and many planters in Bengal retired too. Others moved up the Ganges to the North-Western Provinces (today's Uttar Pradesh) and to Bihar, the last region in British India to sustain indigo plantations. Indian manufacturers, too, increasingly set up in competition with Europeans. At the end of the century, of more than a million and a half acres of land under indigo cultivation throughout India, two-fifths was in northern Bihar.[147] Its production there was as politically, socially and racially controversial as it had been in Bengal.[148] An account of the industry in 1899 states that it employed a million and a half people, seven hundred of them Europeans, with a capital investment of about five million pounds sterling.[149]

Among the other Indian suppliers of indigo, Madras produced a large amount, especially in the second half of the nineteenth century.[150] The EIC had not promoted the Bengali system here for fear of overproduction, instead encouraging local people to produce indigo on its behalf. This way everyone could profit, although quality control was a constant headache, as it was with the Ceylon product.[151] The French, too, produced much indigo in southern India. According to one late nineteenth-century source the production of 'Madras' indigo around

Pondicherry at that time took up nearly 25,000 acres of land. Much of this was consumed locally to dye the blue *guinées* required, as we have seen, for commerce in Senegal and other parts of West Africa.

Figures for total annual indigo exports from India between 1877 and 1897 show a general rise, reaching a peak in 1896 of nearly 10,000 tons, worth over three million pounds sterling in those days.[152] De Lanessan writes of no less than forty-three varieties of Indian indigo on the market – the best being the ones with a reddish/purplish tinge, their quality verifiable by taste and feel.[153]

In Britain trade in indigo, and other natural dyestuffs like fustic and camwood, was still considerable between the two World Wars. London's last drysalter to trade in natural indigo for commercial use sold his final chest of Bengali indigo soon after the Second World War.[154] Natural indigo was still in demand elsewhere as well as for the local consumption, and for this reason India still had almost 11,000 acres of land under indigo cultivation in 1955–6, producing 130 tons of dyestuff.[155] At the end of this century of change, just a very few farms in southern India keep the torch of indigo production alight.

Java in the nineteenth century

Despite India's pre-eminence, there were many other sources of natural indigo in the nineteenth century. In the 1840s a Frenchman, Persoz, listed them as: from the East – Bengal, Manilla, Madras, Coromandel, Java and Bombay; from South America – Caracas and Brazil; from 'North' America – Guatemala and Carolina; and from Africa – Egypt, Senegal and the Ile de France (Mauritius). He declares the principal suppliers to be Bengal, Java and Guatemala,[156] and describes the various shapes and sizes of the dye blocks and their minutely differentiated qualities: the factories of Mulhouse in France were supplied with fifteen kinds from Bengal, each separately priced, and no less than twenty-one from Java.

The Dutch derived great profit from Javan indigo, especially after the withdrawal of their English rivals in 1683, until the early years of the twentieth century.[157] However, there was a brief interruption in 1811, when Java came into British hands following its capture from the Dutch who were at that time under French rule. Sir Thomas Stamford Raffles was appointed English Governor of Java until 1816, after which, with the Napoleonic Wars over, the island returned to Dutch control. During his brief spell on the island Raffles, a passionate reformer, found time to compile his famous *History of Java*. In this work, highly critical of Dutch colonial rule, he comments, among other things, on indigo's high trade value[158] and the unjust methods used to obtain it. In Java local chiefs were compelled

Sample box of forty types of indigo from India, Java and Guatemala imported by S. Schönlank, Berlin, in 1756.

to make annual deliveries of indigo, rice, pepper, etc. to the Dutch company for little or no return so that, as in India, local producers lost out.[159] At this stage only a few attempts had been made by the Dutch and an enterprising Frenchman to manufacture indigo on the island themselves.[160]

During their occupation of Java, however, the British established indigo plantations which the Dutch nurtured once the island was safely back in their hands.[161] Java's excellent indigo provided them with a highly lucrative trade for the rest of the nineteenth century.[162] At the end of the century Dutch traders were still exporting from Java well over two million pounds, produced on about 150 plantations,[163] but by 1910 the industry had largely given way to sugar.

In an interesting aside in his *History* Raffles comments on the most valuable goods sent home by Chinese immigrants working in Java on ships returning to China. Indigo features in an unusual mix that also includes birds' nests, Malayan camphor, dried sea slug, opium, leather hides, gold and silver.[164]

Blue revolutions

Smouldering unrest among indigo workers who resented both working conditions and being forced to use up land that could produce food crops, was manifest in different nations at different times. Revolutionary leaders exploited such discontent to bring about political and social change, and in several places a connection between indigo and slavery or forced labour lingers on in the collective folk memory.

Indigo's most far-reaching political influence was in India, where the injustices of the plantation system in the north-east were central to awakening a new national consciousness in favour of the rural masses, and to promoting political and social change.[165] A simmering discontent among indigo workers resurfaced during the First World War when German synthetic indigo was cut off and plantations had expanded to take advantage of a world shortage of supplies. The desperate conditions of the peasants came to the notice of Mahatma Gandhi soon after he returned from Africa. In 1917 he went to Champaran in northern Bihar, the scene of indigo rioting in 1868,[166] to hold an independent inquiry into peasant exploitation, much to the fury of the Bihar Planters' Association. He refused to comply with an order to leave the area and was put on trial. However, his defiant yet respectful stand caused the case to be withdrawn. His support of the indigo peasants' cause, which in many ways echoed the sympathetic pleas of the more enlightened British administrators and missionaries in Calcutta in 1860, was Gandhi's first act of peaceful civil disobedience on Indian soil and made him a national hero. Thus did indigo start the Indian nationalist ball rolling along a path that led, after thirty years, to the ending of the British Empire in India.[167]

In Central America declining levels of indigo production (which relied for profit on direct trade with Spain) have been viewed by historians as a principal cause leading to independence.[168] It was in the indigo-growing provinces of Spanish Guatemala that unrest was fomented, as in French Saint Domingue.

As for Africa, one historian has explored the complex social, political and economic tensions which arose in the second half of the nineteenth century out of the increased demand for indigo-dyed clothing in western Africa. In the Middle Niger region, now Mali, Maraka (originally Soninke) people were able to monopolize the expanding indigo industry by leaning heavily on slave labour, which increased locally as trans-Atlantic slave exports declined. In 1905, however, the slaves had had enough. They deserted their masters and established themselves as dyers and weavers elsewhere in the region, setting off a 'radical social movement which over the next ten years spread throughout the Western Sudan [i.e. the Sudanic belt]'[169]

Detail of a late nineteenth-century cotton batik sarong of the north-coast Java type. Traditional Javanese designs are combined with newer pictorial motifs, some reflecting Dutch colonial influence.

Colonel Wingate questioning the Mahdist Amir Mahmud, after his capture in Sudan in 1898. Amir Mahmud is wearing the smart patched jacket, *jubba*.

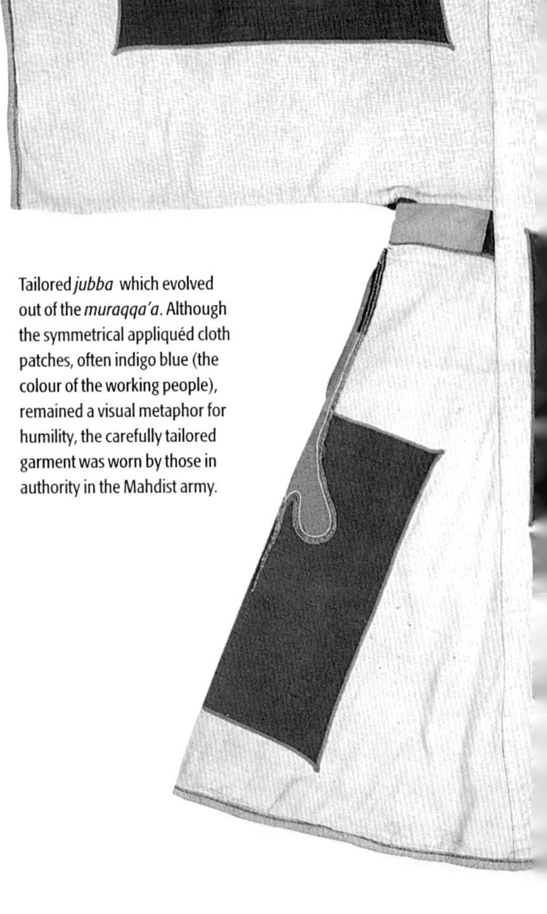

Tailored *jubba* which evolved out of the *muraqqa'a*. Although the symmetrical appliquéd cloth patches, often indigo blue (the colour of the working people), remained a visual metaphor for humility, the carefully tailored garment was worn by those in authority in the Mahdist army.

Patched and ragged tunic, *muraqqa'a*, worn by the first followers of the Mahdi in Sudan, the 'dervishes', as a sign of poverty and humility. Such tunics had for centuries been the dress of Sufi religious orders.

Elsewhere in Africa indigo made a bad name for itself. In nineteenth-century Egypt and Egyptian Sudan monopolies imposed by the Ottoman authorities necessitated intensive production of cash crops – wheat, indigo, cotton and sugar – and this did lasting damage to riverain societies. In the 1980s farmers in Upper Egyptian oases still associated indigo cultivation with oppressive Ottoman rule. A Sudanese saying, 'Where a Turk treads, no grass will grow', expresses a similar sentiment. In the Berber province of Sudan the resentment caused has been seen as contributing to its early fall into the hands of the rebellious Mahdi Mohammed Ahmed and his followers in 1882.[170] Incidentally, the Mahdi's original adherents deliberately wore clothing patched with indigo cloth as a symbol of religious asceticism and humility.[171] The anti-establishment metaphor expressed visually in these indigo patches was reasserted in a different context almost a century later by the adoption of torn and patched blue jeans as the uniform of youth all over the world.

Natural versus synthetic indigo

To replace a manufacture depending on an interesting organic process carried on under healthy conditions in the open air, a manufacture which brings wealth into poor districts, and introduces system and order and civilisation among uncultured people, by one carried on perchance in some dingy sepulchral cave in a chemical works by some fixed and unalterable process, might . . . be a doubtful advantage. (H.E. Schunk, 1897)[172]

A British report on the indigo industry of Bihar in the closing years of the nineteenth century, when the market for natural indigo had reached a peak in Europe, concluded ominously: 'From a scientific point of view, the production of artificial indigo is undoubtedly a grand achievement, but if it can be produced in large quantities at such a price as to render indigo planting altogether unprofitable it can only be regarded as a national calamity.'[173]

It had taken chemists almost fifty years of determined research to reach this critical point for natural indigo. From 1856, thanks to an accidental discovery by the English chemist, William Perkin (who was actually trying to synthesize quinine to combat malaria in the British army), aniline dyes based on coal tar were rapidly developed. Before long various mauves, reds and yellows were being produced synthetically, and the commercial production of synthetic alizarin by German chemists was a real success story.[174] It devastated the French and Dutch madder industry in the 1870s. Immense efforts were made to repeat this achievement with indigo blue but true to form it proved a much harder nut to crack. In 1865 the Nobel prize-winning German chemist Adolf von Baeyer had already worked out its basic chemical structure. He took up the challenge to synthesize it, supported by Heinrich Caro, head of research at the Badische Anilin Soda Fabrik (BASF). It was not, however, until 1897 that the company at last succeeded in launching onto the market the synthetic product, known as 'Indigo Pure',[175] having spent eighteen million gold marks on research, more than the capital value of the company at that time.[176] In 1899 a scientific report declared: 'The stake for which the German chemical manufacturer is playing is a very big one – far bigger than was offered by madder thirty years ago.'[177] But the gamble paid off dramatically. Furthermore, the changes introduced into the chemical industry by BASF in the course of its indigo experiments greatly influenced the development of other German industries, including military ones.[178]

Letter from the university professor Adolf von Baeyer to his friend Heinrich Caro, head of research at BASF, Ludwigshafen, in August 1883. In the midst of private matters the letter communicates the formula for indigo.

Since 1897 the synthetic indigo industry has largely been monopolized by a few chemical giants. Initially German, French and Swiss firms competed for the market, causing price cuts which further undermined natural indigo.[179] In Britain a German firm, taking advantage of complex patent laws to try and corner the British market, started in 1909 to make

Indigo reseach laboratory of the BASF company around 1900.

indigo at Ellesmere Port near Manchester.[180] They used German raw material which became unobtainable when the First World War started. The German staff were interned, but not before they had destroyed vital documents relating to their secret process of indigo manufacture. However, in 1916 Levinstein Ltd, a local dyestuffs' firm, took over the works and the missing technical knowledge was soon reacquired. Indigo production again got under way at Ellesmere Port, but this time as a war exigency to satisfy the enormous demand for the dyestuff to colour service uniforms. (The Swiss, too, greatly increased their indigo output during the war.) After the war the British government was determined not to return to the pre-war position of reliance on a powerful German chemical industry. This led to a merger of Levinstein with British Dyes Ltd in 1919, to form the British Dyestuffs Corporation (BDC). And this in turn amalgamated with other companies in 1926 to become ICI, a big producer of indigo. In 1993, with the demerger of ICI, its Ellesmere Port factory became part of Zeneca plc, and in 1996 Zeneca's textile dyes business was sold to BASF, indigo's original German manufacturers a century before! In 1997 BASF produced 7,000 tons of the annual global total of 20,000 tons of synthetic indigo, Buffalo in the USA 3,500 tons, Bann of Brazil 2,000 and the rest was made in Japan and China. Although synthetic indigo in whatever form is relatively cheap to make, the high demand and the control of production by so few companies keep prices high. Recently, a pre-reduced liquid indigo, which is 'greener' (in environmental terms), has been launched on the denim industry (see p.232).[181]

In the early days synthetic and natural indigo ran side by side. If one puts together the jigsaw of innumerable short reports in chemistry journals either side of 1900, a mixed picture of alarm and hope builds up: hopeful businessmen in the German chemical industry; worried traders and planters in India, Java and elsewhere; and disagreements among dyers, some for and some against the new product. As with the arrival of tropical indigo when it threatened woad, misinformation was spread by interested parties on both sides as to the relative merits of the two forms of the dye.

Initially, there was room for both indigos. In 1900, for example, even Germany was importing 1000 tons of natural indigo from India, despite being the producer of synthetic indigo. In the Netherlands a report of 1902 states that 'in spite of the threatening competition of the synthetic article, there is still an opening for the colonial product' (some 3,350 chests were exported from Java in 1901).[182] In that same year, however, the German Ministry of War decreed that army uniforms should be dyed with German synthetic indigo as the colour was better,[183] while the French government was insisting that for dyeing uniforms it was essential to use natural indigo, which contained 'resin', suitable for cloth 'that is to stand much exposure to the weather'.[184] The British, too, stipulated natural indigo for service uniforms to protect its Indian industry. British planters in India quickly rallied to form the 'Indigo Defence Association', dedicated to 'taking time by the forelock' to save the 'vast interests of our Indian Empire'.[185] With heavy irony, one British observer in India expressed sympathy with the plight of the indigo-planting community, which had 'adapted itself to a life of sport with frequent visits to England and well-lined pockets'.[186] To place the industry on a more scientific footing, research stations were set up in Bihar.[187] Despite this, of the 2,800 large factories (and 6,000 smaller works) producing indigo in the 1880s, by 1911 only 121 remained, and by 1914 competition from synthetic indigo had forced prices of natural down by more than 50 per cent.[188] But the First World War afforded the planters a welcome reprieve. Not only India but other indigo-producing countries in the East stepped up production to stop the gap caused by the cutting off of German synthetic indigo supplies.[189] Even in 1918 Davis, indigo research chemist to the Government of India, was suggesting that the industry could be saved by growing 'Java' indigo and streamlining production methods and organization.[190]

The actual blue colour of synthetic indigo is identical to the natural product. However, indigo extracted from plants contains various so-called 'impurities' such as 'indigo red', i.e. indirubin, which some dyers believe can make the colour more interesting (see p.132 and Appendix). In the prolonged indigo debate that took place among chemists when synthetic indigo first appeared, protagonists of natural indigo pinned their hopes on such 'impurities'. A fascinating scientific report of 1899 asserts:

Very many dyehouses where Indigo Pure *has been tried, have again returned to the natural product, in consequence of the faults … occasioned by the great purity of the artificial product. BASF attributes the advantageous results in the case of natural Indigo specially to indigo gluten, and in their French patent … direct the addition of substances as soap, starch, glue, casein etc. to the vat, in order to obtain the same effect with* Indigo Pure.

The report's author disagrees about the gluten but maintains instead that the presence of indirubin in plant indigo is essential to achieving a satisfying deep blue. He and other chemists therefore rejected forecasts that plant indigo would suffer the same fate as had madder when its rival synthetic alizarin came along.[191]

Various disparate factors affected the speed at which synthetic indigo took over from the plant product in the world as a whole, some relating to cost and availability, particularly in remote regions, but some being

BASF indigo factory before 1930. In those days the oxidized indigo was collected using filter presses, work that demanded hard physical labour.

political. In the West the Second World War in this respect had a similar effect to the First, although on a much smaller scale, hence the fact that Bengal indigo was still commercially traded in London in the early 1950s. When synthetic indigo reached societies that were technologically underdeveloped, it was often incorrectly used (echoing the seventeenth-century European experience with imported indigo), thus reinforcing a preference for the natural product. People were also by nature resistant to new dyestuffs due to deeply held beliefs about the nature of natural dyes, some of which are still held today. In fact, BASF even had to add artificial malodorants to make their product more appealing to dyers who missed the characteristic smell of plant indigo! In Asia, Africa and Latin America local plant indigo was widely used throughout the first half of the twentieth century in many societies, but today synthetic indigo has generally been accepted, or it is added by local dyers into a natural indigo vat, rather as indigo was mixed into the woad vats when it first arrived in Europe. Plant indigo, however, holds its own in traditional medicine and in dyeing related to religious ceremonies where only the natural product will do.

A pair of Levi's waist overalls, thought to be the oldest pair of 'jeans' in existence, *c.*1890.

Other social and economic factors have hastened natural indigo's demise. As well as changing agricultural patterns, the general globalization of societies, the availability of Western clothing and factory goods in places formerly dominated by indigo-blue fabrics, changing tastes in fashion, and the urban aspirations of the young have all ensured that natural indigo will never return to its former position.

As for synthetic indigo, its own future was looking decidedly wobbly by the middle of the century as there were so many other artificial blues on the market, notably Indanthren Blue with its superior fastness to washing and rubbing. Moreover, Mao Tse-tung's banning of imported indigo, hitherto used to dye the famous 'Mao suit' universally worn in China, further depressed the market, as China (like Japan and the USA) had until then been an enormous consumer.[192] Synthetic indigo was going the way of its plant predecessor when it was saved by the astonishing blue jeans revolution which followed the Second World War. In the manufacture of jeans indigo's inherent weaknesses become its strengths, for no other dye can create the special look of blue denim. In the mid-1960s demand for indigo soared, wrong-footing the industry just when it was poised to phase out production. Who could have foreseen that denim would capture and maintain a loyal and truly universal clientele?
A century after it was first launched on the commercial market, synthetic indigo is in constant demand, mainly to dye sufficient yarn to fulfil an annual global output of more than a billion pairs of jeans![193]

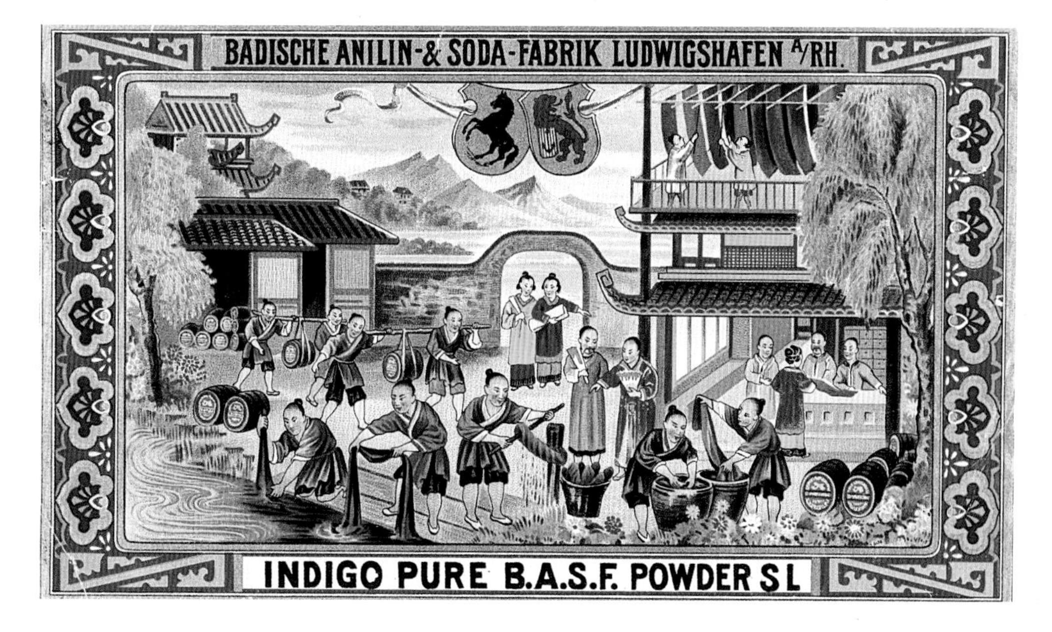

Label used in 1903 for indigo shipments to China, at that time one of the most important export markets for BASF synthetic indigo.

Of all productions called colonial,
indigo is the one which demands, in the manufacture,
the largest share of intelligence and judgement
(JOHN CRAWFURD, 1820)

Indigo Plants
and the Making of their Dye

Without indigo the world would have had no colourfast blue dye until the nineteenth century. Luckily, indigo was more widely available than any other natural dye as it can be obtained from species of many different plant genera found in several parts of the world (see map, p.90). 'Indican', which is the chemical source, or 'precursor', of the dye and is present in the leaves of all the indigo-bearing plants (see the Appendix), produces the same indigo-blue colourant after processing whatever plant it comes from. Each species has its own distinctive characteristics and production methods, but the chemical make-up of their actual dyestuff is common to all. Whether indigo's plant source is woad from a Thuringian farmstead, an indigofera from India, Africa or Central America, *Polygonum tinctorium* from a Japanese island, or species from one of the other genera grown by villagers in the Far East or West Africa, the resulting blue is indistinguishable even to the specialist.

The reason that, perhaps confusingly, one genus came to be named *Indigofera* – i.e. 'indigo bearer' – by Carl Linnaeus (1701–78) when founding the modern botanical classification, is that species within this genus provided most of the dyestuff used in inter-continental commerce. However, other genera have also been widely used since antiquity, and these will be examined before describing cultivation methods and the difficult processes of extracting indigo from the plant, on both local and commercial scales.

Chinese village dyer with indigo paste made from the plant *Strobilanthes flaccidifolius* (resting on the basket). She is wearing indigo-dyed clothing, and behind are old aprons re-dipped in the indigo vat to freshen them up. Near Kaili, Guizhou province, 1993.

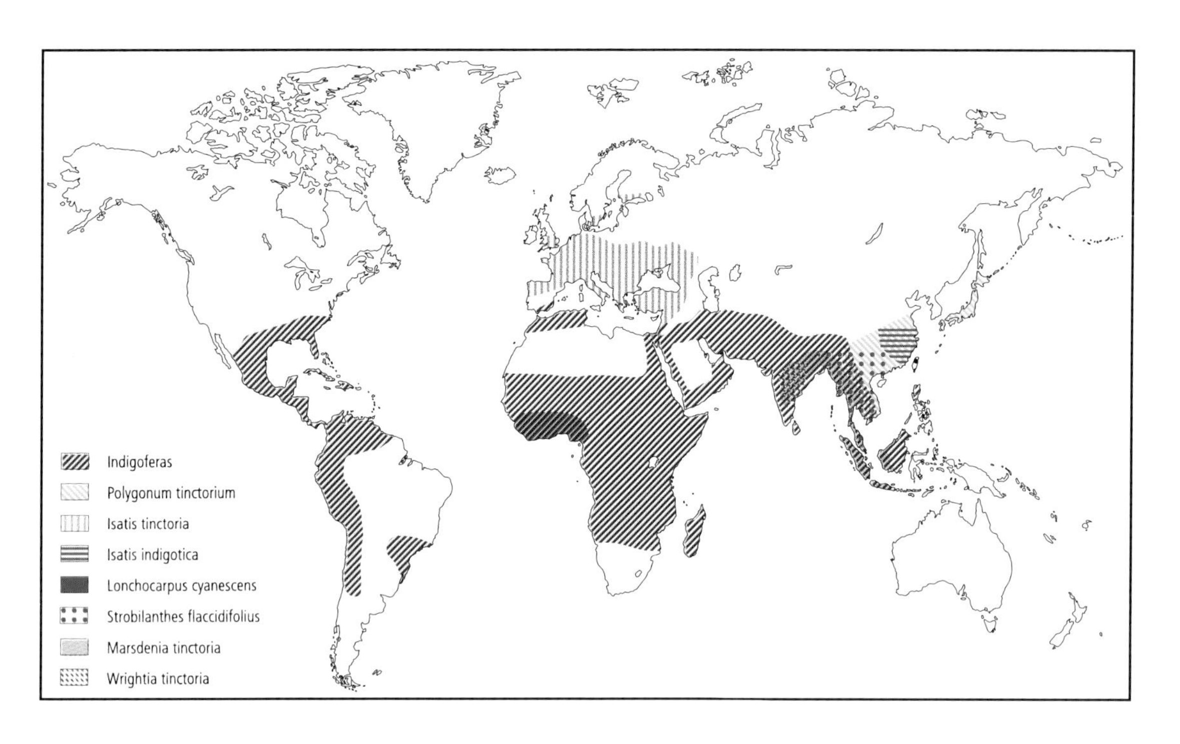

Map showing distribution of main indigo-producing plants.

There has been considerable dissent among scholars about the diffusion and domestication of indigo plants. Evidence points to the fact that the colouring ability of indigenous indigo plants was discovered independently by people in different continents, although indigo species and dye production techniques were later deliberately exchanged between countries and continents for commercial purposes; and India was a main focus of this enterprise. In colonial times it was common to find indigofera spreading hand in hand with sugar-cane and cotton, which require similar growing conditions. But as regards the spread of uncultivated indigo species in prehistoric times, the lens is much cloudier. In the case of the Middle East, for example, there has been a general assumption of a diffusion from the East. However, recent research by palaeobotanists suggests that indigofera species could originally have spread eastwards from Africa to Arabia and India as pan-tropical weeds along with cotton and cereal.[1]

Now that techniques of scientific analysis are becoming more refined and interest in the subject is growing, botanists and organic chemists are conducting ongoing research into types and properties of dye plants. Botanical identification of dye-bearing indigo species can be a particular minefield for the non-specialist. It is helped neither by the fact that so many species were transferred between countries before the Linnaean system was introduced, nor by the fact that certain species have been misidentified in the past. Furthermore, some genera have been divided and renamed by botanists in recent years. To save unnecessary confusion, long-standing familiar names will be used here, with any new classification noted in parentheses after the initial mention of the plant.

Indigofera tinctoria cultivated at Ba Saang Sa, village weavers' project, near Nong Khai, north-east Thailand, 1991.

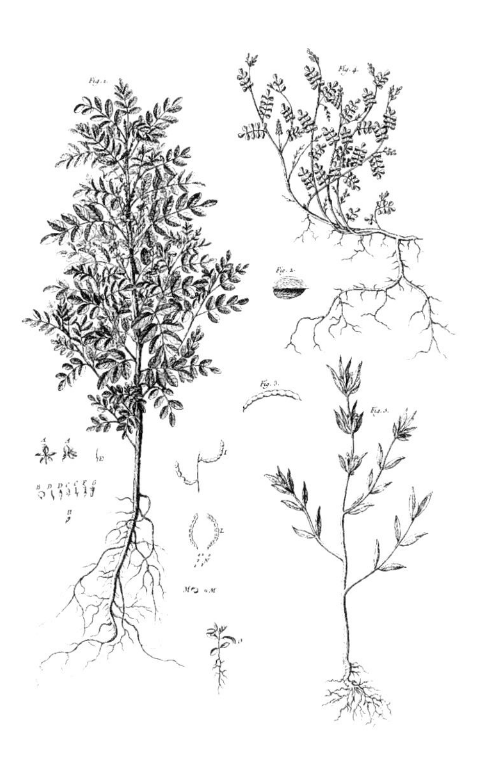

Eighteenth-century French engraving of indigofera plants.

The indigo plants[2]

Indigofera

The genus *Indigofera*, the third largest in the family Leguminosae, consists of almost 800 species. These species can grow on land between sea level and 1650 metres. Over 600 can be found in Africa,[3] nearly 200 in Asia, about 80 in America and 50–60 in Australia.[4] No one can yet explain why only a handful of these should have a high content of indican, indigo's precursor. The existence within the overall genus of a section also called *Indigofera,* which contains the dye plants, may have contributed to confusions over nomenclature.

The most widely exploited of all indigo plants is the species *Indigofera tinctoria* L. (= *I. sumatrana* Gaertn.), a perennial shrub which has been cultivated in most tropical and subtropical regions. It is thought to have spread with dyeing technology from India first to Southeast Asia,[5] then through the Middle East to parts of Africa,[6] including Madagascar, and later to America. (This is indicated by the use in these places of words based on the Sanskrit for indigo, *nil*).[7] Under favourable conditions *I. tinctoria* can grow nearly two metres tall. It is distinguishable from the other indigo species by its comparatively large paired leaves, not unlike rue leaves, and its long, relatively thin, straight or arc-shaped pods.

The other main indigoferas used for dyeing were the African *I. arrecta* Hochst. ex A. Rich and the similar looking *I. articulata* Gouan and *I. coerulea* Roxb. These often have leaves with more numerous and smaller leaflets than *I. tinctoria* and differently shaped pods (straight and relatively thick in *I. arrecta* and strongly beaded in the others). *I. arrecta*, widespread in Africa, was introduced to Java in the middle of the nineteenth century, where it was called 'Natal' indigo. It was later cultivated in India, where it was known as 'Java' indigo. It was also grown elsewhere in Indonesia (Sumatra, Sumba and Flores), and in Vietnam, Laos, the Philippines and the Near East. It has larger leaves than *I. tinctoria* and was considered

the best indigofera dye variety. *I. articulata* and *I. coerulea*, both often referred to erroneously as *I. argentea* L., were widely cultivated in northwestern India, the Arab world and West Africa, being more suited to drier climates than *I. tinctoria*.[8] In the golden days of Islamic textile production indigoferas were also cultivated in the Mediterranean islands of Malta,[9] Sicily and Cyprus, as well as in Spain.[10]

The other major indigos of commerce were the native tropical American species, *I. suffruticosa* Mill. (= *I. anil* L.) and *I. micheliana* Rose (= *I. guatemalensis* Moc., Sessé and Cerv. ex Backer), sometimes mistakenly considered a subspecies of *I. suffruticosa*.[11] The former was also cultivated in North America, introduced by the Dutch to Southeast Asia and adopted in China and Africa.[12] As with the Asian species, the diffusion of American species has been the focus of much botanical study, as there has been considerable confusion about the species used in commerce. It is thought that dye species dispersed from western Mexico to the rest of Central and South America in both pre- and post-Hispanic times and that *I. suffruticosa* and *I. micheliana* are native species, while *I. tinctoria* was a post-Conquest introduction.[13] *I. caroliniana* Mill., a native North American species, was also used commercially in the southern states of North America.[14]

Indigoferas' cylindrical seed pods range variously from straight to crescent shapes, depending on the species. Not mincing his words, the herbalist Parkinson in 1640 described them as 'hanging downewards, like unto the wormes ... which we call *arseworms*, yet somewhat thicke and full of blacke seede'.[15] The small seeds have been compared to gunpowder by many writers.

The supremacy of indigo-bearing indigoferas over any of the other contenders is not related to the quality or quantity of the dyestuff produced but to the distribution of the species with their suitability for cultivation in so many European colonies, and the practice of drying the dyestuff extracted from the leaves. Concentrating the dye in a durable state had obvious implications for trade. Many of the other major indigo plants have traditionally been made into a wet paste or leaf compost, serviceable for short distance trade and easier to use, but impracticable for intercontinental commerce. Of these plants, the two most important were the temperate woad, *Isatis tinctoria* L., and *Polygonum tinctorium* Ait. (= *Persicaria tinctoria* (Ait.) H. Gross), the main source in China and Japan. Other indigo plants of more localized importance include *Lonchocarpus cyanescens* Benth., *Strobilanthes flaccidifolius* Nees (= *Baphicacanthus cusia* (Nees) Bremek; = *Strobilanthes cusia* (Nees) Kuntze), *Marsdenia tinctoria* Roxb. (= *Asclepias tinctoria* Roxb.), and *Wrightia tinctoria* (Rottler) R. Br. (= *Nerium tinctorium* Roxb.).

First year woad plants rea[dy] for dyeing in early summe[r], Devon, 1996.

Illustration of the biennia[l] woad plant, *Isatis tinctor[ia]*, second year, with flowers and seeds. From Schrebe[r's] famous book on woad, published in 1752.

Isatis tinctoria

Isatis tinctoria, or woad (often known by its French name, *pastel*), is one of over fifty species belonging to the genus Isatis. It is a biennial member of the family Cruciferae, related to the common cabbage.[16] Native to the Mediterranean and western Asia, it is not known when it spread to north-west Europe, although it had reached Britain by the Iron Age and it is found as far north as Scandinavia. Woad was also grown in the Azores and taken to North America by the early settlers. In northern Europe it tends now to be found only in places where it has escaped from former cultivation. The first year plants, which alone contain the dyestuff, resemble rosettes of spinach leaves. In early summer of the second year, when the flowering spikes shoot up to well over a metre, the distinctive arrow-shaped leaves embrace the stem, and there is a mass of tiny yellow flowers in the branching head. In medieval Europe fields of flowering woad (grown for its seed) would have been as distinctive a landscape feature as are crops of its relations, mustard and rape, today. The large pendulous dark seed pods, aptly described by Gerard in his herbal as 'like little blackish toongs [tongues]',[17] are also striking.

Although the main species to have been exploited commercially is *Isatis tinctoria*, many other strains and species have been found to contain some dyestuff.[18] One of these was reported to be grown in Spain and Portugal,[19] while in Turkey botanists have identified over thirty-six species and subspecies, some of which are being used today to provide dye for modern naturally dyed rugs. Recent research has revealed that some Turkish dye-bearing species of woad contain mostly indirubin, the 'indigo red' component normally present only in small proportions in indigo dye plants, and even that colours other than blues can be produced from them by different dyeing processes.[20]

In China a further woad species, *Isatis indigotica* Fortune ex Linl. (known locally as *sung lan* or *tien-ching*, and in English as 'tea indigo' or 'cabbage blue'), was identified by the botanist Robert Fortune in the 1840s. This species, which appears to have a higher dye content than other woads, had probably been introduced in the sixteenth century for use in areas too cold for tropical and subtropical indigo plants.[21] Fortune found it being cultivated in abundance around Nanking to supply dyers of Shanghai and other northern towns,[22] while a seventeenth-century Chinese source mentions its cultivation on the hillsides of Fukien province to the south.[23] Watt later mentions in his economic dictionary that woad or 'devil's weed' was also to be found both growing wild and cultivated in Tibet and Afghanistan.[24]

Although enormously important for the medieval European woollen industry, woad was inevitably elbowed out by imported indigo dyestuff when this became freely available, just as American cochineal would usurp the Mediterranean kermes insect red dye. Although the same blue dye substance is produced from all the indigo plants, far more indigo was extracted from the tropical and subtropical species than from woad. Moreover, the dry, concentrated indigo produced from the former for the export trade was dissolved in vats of high alkalinity, making it compatible with cellulose fibres like cotton and flax, whereas woad dyestuff only worked well with woollen fibres.

Polygonum tinctorium left to flower for seed for the following year by indigo farmer Osamu Nii. Tokushima, Japan, 1993.

Polygonum tinctorium

Polygonum tinctorium, an annual or biennial herbaceous plant known as 'dyer's knotweed', 'Chinese indigo', or 'Japanese indigo', belongs to the large family Polygonaceae. Although several species of Polygonum have been noted as bearing indigo, they may in fact all belong to the *P. tinctorium* group. Plants reach about half a metre high, and have large dark bluish-green leaves alternating up the fleshy stems, whose colours vary from green to almost red. Its small flowers range in shade, too, from white to dark pink. Some people believe that the pink-flowered varieties produce more dye than the white. When manufactured by traditional methods similar to those formerly used with woad, it produces far more indigo dye than the same quantity of woad leaves.

P. tinctorium was described by the late China scholar Joseph Needham as 'the ancient and indigenous blue dye-plant of China, the father and mother of all those millions of good blue garments that those who have lived in China know so well'.[25] It is still grown there, being known as *liao-lan*, while *Indigofera tinctoria* is *mu-lan*. Indeed the Chinese character for blue is the same as that for indigo plants. As *P. tinctorium* is subtropical, it could be cultivated in China both alongside indigofera and in places too cold for the latter. It is thought to have been introduced into Japan, where it became known as *Ai*, from southern China sometime after the fifth century AD. It became the main dye plant of Japan and was widely cultivated there. Throughout the countryside plants were grown at home for village dyers, while places like Tokushima district made indigo on a commercial scale. It has also been widely cultivated in Korea and Vietnam.[26] Europe became interested in 'Chinese indigo' at the end of the eighteenth century as it can thrive successfully there out of doors if frost damage can be avoided. On the continent various countries, notably France, made concerted efforts to grow their own plants, while in England dyers tried out the dye itself, imported as 'Persicaria'.[27] These various trials were on the whole successful but short-lived. New trials have recently been undertaken in England.[28]

Japanese indigo, *Polygonum tinctorium*, grown in southern England by the author.

Adding fresh leaves of *Strobilanthes flaccidifolius* to a wooden barrel to make indigo dye. Zhaoxing village, Guizhou (Dong minority area), south-west China, 1993.

Strobilanthes flaccidifolius

A perennial sub-shrub of the family Acanthaceae, *Strobilanthes flaccidifolius*, known as 'Assam Indigo' (or locally as *rum*), has been a source of indigo in mountainous regions of Asia, notably in central and south-west China,[29] Thailand, Burma, north-east India, Bhutan, Laos, Vietnam, Bangladesh and Malaya.[30] It is often the only natural dye still used by people of various ethnic minorities living in remote regions of these countries. It was also cultivated in Taiwan and the southern Japanese Ryukyu islands, notably Okinawa, where it is still found. In the nineteenth century Watt urged indigo planters in India to grow *S. flaccidifolius* to supplement indigofera, arguing that it could be cropped in the season when the indigo factories lay idle.[31]

S. flaccidifolius plants reach a metre high. They have mauve, trumpet-shaped flowers when left to bloom, but when used for dyeing, they are propagated annually from cuttings and harvested before flowering, or the young leaves are collected from wild plants. The dye content is high[32] and the plant appears to produce particularly dark shades.[33]

Lonchocarpus cyanescens

A widespread source of indigo in much of West Africa is the native legume *Lonchocarpus cyanescens*, sometimes known as 'Yoruba indigo' (or *elu* in Yoruba). It was also introduced into Malaysia.[34] It is a fast-growing woody liana, reaching well over three metres high, with large leaves and panicles of pea-like purplish flowers. When cut down it soon re-grows, and, like indigofera, young plants produce the best dye. Its dye, as well as that from indigofera (probably introduced with Islam) is generally known locally as *gara*. In the early nineteenth century the French explorer René Caillié noted dyers using both plants in the northern Ivory Coast and in Sierra Leone.[35]

Marsdenia tinctoria

Another rampant vine, *Marsdenia tinctoria* (family Asclepiadaceae), was widely used in parts of Southeast Asia, where its commonest name is *tarum-akar*, as an alternative source of indigo. It was named after William Marsden, who collected specimens in Sumatra, hoping they might be of commercial value in Britain's colonies. He presented them on his return in 1780 to the botanist Joseph Banks.[36] The plant with its long dark oval leaves and round clusters of small yellow flowers is found in regions stretching from the north-east Himalayas and Burma to the islands of Indonesia, and was also cultivated in the Indian Deccan. It could be harvested throughout the year and grows in regions too wet for indigoferas.[37] It was used either fresh in the dye vat, or treated like indigoferas to produce dyestuff for trading.

Wrightia tinctoria

One other significant indigo dye plant is *Wrightia tinctoria* (family Apocynaceae), known by its Latin name *nerium*, 'dyer's oleander' or 'Manila indigo'. A small, fine-branched tree, it has narrow oval leaves with large heads of scented white flowers and pendulous seed pods. It grows naturally in central and southern India, Burma and Malaya. Used by the people of the Deccan, it was commercialized in the Madras region, where it was known as 'Pala indigo'. In the 1790s the British explored its commercial potential,[38] Roxburgh's account of its use in India being published in 1811.[39] The process of dye extraction was the same as for indigoferas, except that hot water was required (as it is for modern woad extraction).

Lesser, or 'false' indigo plants

In addition to the above, other plants have been said to produce indigo. According to Watt, *Tephrosia tinctoria* Pers. and *T. purpurea* Roxb. (Leguminosae) of central India, were used for indigo dyeing, as was another species found in Egypt.[40] Furthermore, there is a strange group of plants that have been reputed to produce either genuine indigo or 'false' (or 'pseudo') indigo. Among these is *Mercurialis leiocarpa* Sieb. and Zucc. (Euphorbiaceae), which is closely related to dog's mercury and is found in India, China and Japan, where it is known as *yama ai*.[41] Other plants in this group include *Desmodium brachypodum* A. Gray, which was used for unusual bark-cloth dyeing on the Solomon islands;[42] *Scabiosa succisa* L.; and *Baptisia tinctoria* Roxb. (= *Sophora tinctoria* L.), which was known as 'yellow wild indigo', 'false/bastard indigo' or 'rattle-weed' and was apparently used both for yellow and blue dyeing in America.[43] Tantalizing footnotes hint at the dyeing qualities of members of this group, but more research is needed. Many of them are discussed by Cardon[44] and listed in the exhibition catalogue *Sublime Indigo*.[45]

Lonchocarpus cyanescens growing wild in Nigeria (see p.95).

Growing conditions and cultivation of indigo plants

Nowadays, when cultivation of indigo plants is unusual, it is hard to imagine the scene centuries past when they were often cultivated as a staple agricultural crop. In China indigo is, with rice, one of the first crops for which there is evidence of transplanting, dating to the second century AD.[46] Much later, as the pressure on fertile land intensified, an indigo crop could be slotted in between two of rice in an annual rotation. This would provide enough dye for local consumption plus a surplus for sale. Today plots of various indigo plants, the last vestiges of China's long history of indigo production, can still be found in minority tribal areas in the mountainous south-west of the country.[47] Similarly in other parts of Southeast Asia and Japan where it was formerly common, it is now only found in the most isolated areas. In north-eastern and southern India (Andhra Pradesh)[48] and in a few parts of West Africa indigo is still grown, while in areas formerly renowned for its cultivation in the Middle East, North Africa and Central America it has almost entirely died out.[49] Apart from the complex social reasons for its rapid abandonment almost everywhere since the 1960s, from an agricultural point of view changing farming patterns, pressure on land and periodic droughts have all contributed.

In places where climate and conditions were ideal, such as on the most fertile of the Indonesian islands or in Nigeria's Yorubaland, excellent dye could be produced from indigo plants growing wild. But cultivating indigo to produce top quality commercial dye required constant vigilance at every stage and in the colonial system was extremely labour-intensive.[50]

Fresh seed, preferably from outside sources, was needed to maintain dye quality. The best fields would be reserved for seed production, or planters would exchange seeds, as Raffles recorded in Java.[51] Colonial planters experimented with imported seed, especially when synthetic indigo threatened the market. Planters in India favoured *I. arrecta* grown specially in Africa rather than *I. tinctoria*.[52] A biologist employed by the Bihar Indigo Planters' Association from 1901 to 1904 describes, with rather too much relish, two ingenious experiments to improve the germination percentage of such 'Natal-Java' seeds:

The first attempt was to feed the seed to ducks and collect it as ejected. This proved quite ineffective because the time of passage was too short. The next, and sufficiently successful, attempt was to mix the seed with emery powder so graded that subsequent separation was easy. The mixture was then . . . spread on a smooth concrete floor and jumped on by chokras *(boys). It was wonderful to see what feet, which had never worn a covering since birth, could endure.*[53]

Indigo dyer in San, Mali, with a gourd full of pounded and dried indigofera leaves, 1997. She is wearing a traditional bride's wedding shawl, made of alternating bands of plain and tie-dyed indigo cloth sewn together after dyeing.

In addition to differences between species, small variations within species such as the size of the leaves or, in woad's case, the hairiness of the leaves could affect dye quality. Even today, where natural indigo is being revived, experimentation to isolate the best species and strains continues.

In naturally fertile and humid areas like Java planters still sought to improve yields.[54] Dye quality was affected by many factors, and in drier or erratic climates much effort was required to prepare the ground and irrigate and nurture the plants. In a bad year ground would have to be re-sown up to four times to produce a successful crop.[55] One laborious and strange system in Bihar involved compacting the soil and sowing seed in the middle of the dry season.[56]

Indigo plants of all varieties produced the best dye from the first harvest of the season, and in most places where indigo was culturally important separate terms were applied to each successive cutting. Furthermore, although indigoferas are perennial, dye quality deteriorated after the third year, and stock needed renewing. The English merchant William Finch in India in 1610 considered indigo from second-year plants to be the best and bemoaned dishonest practices: 'Some deceitfully will take of the herbe of all three crops and steepe them all together, hard to be discerned, very knavishly.'[57] Finch's near contemporary Pelsaert, in Shakespearean fashion compared the first cropping to 'a growing lad', the second to 'a man in his vigorous prime' and the third to 'an old decrepit man'![58] Sometimes local lore dictated the agricultural cycle – in Thailand, for example, indigo plants had to be grown before the bullfrogs began to cry.[59]

Both woad and indigo plants other than indigoferas produce far more dye on ground enriched with nitrogen. Woad, like *Strobilanthes flaccidifolius*[60] and *Polygonum tinctorium*, is greedy for nitrogen, whereas indigoferas fix their own from the air, but grow better with added phosphate.[61] These characteristics affected the histories of the plants. In medieval times deep-rooted woad was the ideal crop to break new ground, but unless the land was given many years' rest after three of growing woad, soil became badly depleted, which sometimes gave woad a bad name. 'Woad sickness' afflicting land in traditional areas like German Thuringia may have contributed to woad's demise in the late Middle Ages.[62] By contrast leguminous indigofera, being naturally rich in nitrogen, is an ideal crop for rotation cultivation, and even today is sometimes grown purely as a green crop, as well as for cattle fodder. The spent leaf mass, too, after dye has been extracted provides an extremely rich compost for ploughing back into the land.[63] In India it was also used to fuel the boilers used in indigo factories.[64]

Just as the lives of European indigo planters in the East and West Indies revolved around their product and their colonial lifestyle – which from all accounts seems to have involved endless shooting parties during leisure periods[65] – so, too, in Europe 'woad men' or 'waddies' had a self-contained way of life, usually passed on down the generations. Interviews conducted by Norman Wills with members of Britain's last woad communities paint a vivid picture of itinerant family groups, who were reputed to have intermarried as the smell of woad processing was such an occupational hazard, and whose lives revolved around the woad seasons.[66] During the harvest period whole families, supplemented by casual labour, would be out in the fields, singing their special woad songs. In the 1880s one Susanna Peckover who looked forward to this season wrote: 'A very merry time is this woad harvest.'[67] The picture would have been similar wherever in Europe woad was processed.

As was the case with indigo dyeing itself, when things went wrong with the cultivation or manufacture of indigo dyestuff a common male reaction was to blame women. In Egypt, for example, the failure of a crop could be attributed to the passage past the field of a menstruating woman,[68] while in one Chinese province women with flowers in their hair (i.e. all women!) were kept away from the jars when men were making indigo dye.[69] In Indonesian Flores women are forbidden to use coarse language when harvesting indigo, as this offends the soul of the plant, which retaliates by refusing to yield its dye.[70] By contrast, writing about production in the West Indies, Diderot reports on a well-established superstition that only odd numbers of seeds should be planted per hole.[71] All such superstitions reflect the fact that making indigo dye can be a tricky business.

Dye production

So much for growing indigo plants. The real challenge for farmers is extracting the dye from their leaves, stage one of indigo's conjuring trick. Stage two, when the chemical reactions required in stage one have to be reversed, relies on the skills of the dyers, the subject of the next chapter.

Like almost everything to do with indigo, its manufacture is uniquely strange. From the time of Pliny on, unlocking the secrets of indigo's chemistry aroused immense curiosity and awe. When one considers that as late as 1705 a patent was obtained for extracting indigo, 'a blue stone', from mines in Germany, one can see there was a lot to learn.[72] Others thought that indigo was a dusting on the surface of the leaves.[73] Scores of experts in the eighteenth and nineteenth centuries applied their minds to trying to understand exactly what was going on when indigo dye was being formed from the leaves of the plants. In the 1850s Edward Schunk wrote: 'The properties of indigo-blue, which are so peculiar as almost to separate it from all other organic bodies, and to constitute it one *sui generis*, naturally suggest the inquiry, in what form it is contained in the plants and animals from which it is derived.'[74] In 1855 his curiosity was rewarded when he managed to isolate indigo's precursor in woad leaves as a compound which he named 'indican'.[75] This was a breakthrough, but for the rest of the century scientists were still struggling to understand exactly how indican, a colourless glucoside, became converted to indigo (i.e. the blue colouring matter sometimes referred to as 'indigotin').[76] Finally in 1898 chemists discovered that, when indigo plants were soaked in water, enzymic hydrolysis transformed indican into indoxyl ('Indigo white') and glucose, and that, when the liquid containing the indoxyl was vigorously whisked to incorporate oxygen, the indoxyl would convert to indigo.[77] (For the chemical formulae see the Appendix.) Even then, as the chemist Rawson noted, 'notwithstanding the great amount of time that has been devoted to the subject, it is not definitely known in what state of chemical composition the colour principle exists in the leaf of the plant, nor the precise changes which take place during conversion into indigo blue'.[78] Nevertheless, the basic principles were now established, although organic chemists continue to be teased by the minutiae of indigo's quirky transformations.

In the case of woad the tale has a strange twist. For much of this century the precursor of indigo in woad was thought likely to be mainly 'isatan B', not indican as is the case with the tropical and subtropical plants. However, new studies have revealed much of the precursor in woad to be indican after all, only the remainder being 'isatan B'.[79] Meanwhile the tropical and subtropical plants have been found to contain some 'isatan B' themselves.[80] Its molecule is similar to indican but is

Oxidizing liquid with a wooden paddle after soaked indigo plants have been removed from a cistern constructed in Spanish colonial days. The surface foam has oxidized to blue, the liquid below remains greenish. Chalatenango, northern El Salvador, 1996.

unstable and breaks down easily. During hydrolysis 'isatan B' converts to indoxyl and keto-gluconate (an extra carbon) instead of indoxyl and glucose (see the Appendix). This may perhaps explain why steeping woad leaves in merely tepid water, as is done with the tropical or subtropical species, is not effective. Woad was therefore processed in a different way, resulting in leaf compost which contained very little indigo compared with that extracted from tropical plants. From the seventeenth century, however, chemists tried hard to extract indigo from woad leaves, with little success, although French experiments made during the continental blockade (see p.58) came close to achieving the solution. The French realized that woad, unlike other indigo plants (apart from *Wrightia tinctoria*), had to be steeped in hot water.[81] Only recently, though, have scientists discovered that not only must woad leaves steep for just ten minutes at 80°C but the liquid must also cool down rapidly.[82] This is now possible under controlled conditions using modern technology. For future producers of woad this may well be a real breakthrough.

In the past what did all this mean in real life? Processing indigo could be done in three ways: first the simple method, where fresh leaves were put directly in the dye pot; second, that where the leaf mass was processed and fermented but the dye pigment not extracted; and third, the most sophisticated method, where the indigo was fully extracted from the leaf mass.

In the first method, used in many traditional societies, the basic chemical transformations are rolled into one in the dye pot. Fresh leaves and water are put into the pot with ingredients such as ash water or urine; this will render the liquid alkaline and set off fermentation, which gradually reduces the oxygen. This system serves for small-scale dyeing but has many limitations. It can only take place near the site of dye plants at certain times of year, and the dye is weak, requiring endless dippings to produce a dark shade. In some Indonesian communities yarn would be dipped in and out of the dye pot for up to a year. Even in unsophisticated communities, however, methods have also been devised to extract a dye paste suitable for short-term storage.[83]

The second method is to start off the fermentation process by composting the leaves, which are then dried to produce a transportable and storable leaf mass that is comparatively quick and easy to use in the subsequent dye vat. This system was used in various ways for processing woad in Europe, and is still used in Japan with *Polygonum tinctorium* and in West Africa with both *Lonchocarpus cyanescens* and indigoferas. In the case of woad and Japanese indigo the leaf mass was subjected to a protracted fermentation process.

Special clay jars made for indigo manufacture, with a bung hole about 10 cm above the base for draining off liquid to leave indigo paste in the base. Palm frond whisk used to aerate the liquid. Near Ibri, Oman, 1985.

Although woad balls were traded within Europe, they would not have been suitable for inter-continental trade, being too bulky and liable to turn mouldy, as well as too weak to dye cellulose fibres. To be really effective the indigo needs to be extracted from the leaf mass by the third method, a process that rates among the world's oldest and most long-lasting agricultural industries. As we have seen, the Indians had discovered the secret in antiquity, thereby producing a luxury trading commodity. It seems likely that this Indian method influenced practices elsewhere, spreading along with the huge network of trade already discussed. This could explain methods still observable in southern Arabia in the late 1980s and 1990s which resemble almost exactly the traditional pre-colonial Indian model. From the seventeenth century European colonial factories clearly developed out of the Indian prototype. Extracting the blue clay-like sludge and making it evaporate to become what is effectively a hard dry lump of insoluble indigo pigment is laborious, and the dyer is then faced with the harder job of grinding it up before converting it into a dye. However, such indigo was ideal for long-distance trade, being highly concentrated and completely durable. As we saw in the previous chapter, its manufacture became big business.

Clay jars filled with fresh indigofera branches soaking in water beside a *falaj* (water channel). Nizwa, Oman, 1981.

Balling pulped woad leaves, which were piled onto a wooden tray on the 'woad horse' to be transported on the head of a 'waddy' (woadman) to the drying ranges (back of picture). Parson Drove mill, *c.*1900.

Parson Drove woad mill (demolished in 1914), in the English fenlands near Wisbech, Cambridgeshire, *c.*1900. The mill was a temporary thatched structure; the wooden horse-powered rollers for crushing woad leaves (on stone slabs) had metal cutting bars.

Adding water, raking and turning a steaming heap of Japanese indigo, *Polygonum tinctorium*, compost in the process of making *sukumo*. Tokushima, Japan, 1993.

The composting of leaves – woad, and Japanese and African indigo

Europe and Japan are linked by the similar way they have treated their woad and indigo plants respectively. Historically, the Europeans had no alternative but to compost their woad leaves, whereas the Japanese with their *Polygonum tinctorium* could have extracted indigo from its leaves but chose instead the composting method.

A few forlorn relics of Europe's woad industry are still standing. As well as the sites of woad mills already mentioned, buildings formerly used to dry woad balls or 'couch' woad still exist in France and Germany.[84] In England only photographs remain to show the country's last roller and 'couching' houses and drying ranges (some were permanent structures, others temporary) but, more poignantly still, Norman Wills' tape-recordings of the reminiscences of the Fenlands' last 'waddies' provide an unique oral continuity with the Middle Ages.[85]

Processing wood for the dyers was a time-consuming business, as it involved a double fermentation to maximize its low dye content.[86] Crops were harvested up to four or five times during the summer, beginning in May. The fresh leaves were taken straight to the circular mill, where they were ground up by horse-driven (or, later, water-, steam- or oil-driven)[87] rollers – always of stone on the continent, but latterly wooden with metal

cutting bars in England.[88] (In Finland an eighteenth-century woad mill was constructed that could be 'powered' by children!)[89] The resulting woad pulp was drained and moulded by hand on wooden boards into balls the size of cricket balls, which were laid out on racks to dry for several weeks. Woad balls at this stage were known as 'green woad', or *coques/cocaignes* in French. In Germany 'green' woad balls were usually transported to the nearest towns, where specialist woad refiners, put under oath not to divulge their secrets, carried out the second fermentation before selling the final product on to merchants or dyers.[90] If one old account is to be believed, the effects of this second fermentation were all-pervasive:

These balls … being piled in heaps grow sensibly hot, and exhale a Urinary Volatile Salt … This spreading Salt not only extends itself to the place where the Balls are, but fills all the Neighbouring Houses with its Smell, and occasions a sort of Dewy drops to hang on the Wall, Roofs and other Parts of the Houses, which evaporate into Air, if the Volatile Salt be not extracted.[91]

When woad was exported in its 'green' state it would often be treated by 'refiners' or 'wad porters' at the port of entry.[92] Otherwise woad men treated their 'green' woad themselves *in situ* at the mills. Wherever it took place, the second fermentation, or 'couching', was the hardest task of all. 'Couching' involved crushing up the woad balls, either with mallets, or 'rammers',[93] or back in the mill. The resulting mass was spread feet deep on the ground and watered. Over the next six to nine weeks the clay-like mass was repeatedly sprinkled with water and turned as it heated up and steamed like compost. This rather smelly process was known in England as 'silvering', and required considerable experience to ensure that the mass did not overheat and that all parts fermented to an equal degree. Couching was complete when the mixture became dry and mouldy – well 'beavered' – at which point the 'couched woad' (*pastel agranat* in

Child pounding leaves of *Lonchocarpus cyanescens* to make into indigo balls. Near Illorin, Nigeria, 1967.

An offering of rice wine, in a sake bottle with a sprig of cedar in it, is always made to the indigo gods after 'putting the baby to bed', i.e. covering a heap of composting indigo leaves with straw mats to keep them warm. Tokushima, 1993.

French) was packed into large sacks or wooden barrels ready for the dyers. If 'couched woad' was left to mature for several years its dye content was said to double.

In Japan a system similar to Europe's for woad was carried out on both small and large scales, probably from about the tenth century.[94] The more commercialized system, widespread in the past, is still in action in Tokushima prefecture (formerly Awa) on Shikoku island. During the Edo period (1600–1868), when cotton became widely available, indigo production brought great wealth to this fertile region.[95] Its indigo was so valuable that farmers in the eighteenth century who gave away the secrets of production to outsiders were apparently beheaded.[96] In 1903 nearly 40,000 acres were still under indigo cultivation. Today a few farmers, with government support, cultivate about 50 acres of *Polygonum tinctorium*, known as *tade-ai*, to keep this most traditional of Japanese industries alive.

Indigo seeds are planted in beds in late winter, and in April the small plants are transplanted into fields. The leaves are harvested twice in the summer. The Japanese, unlike the woad makers, do not 'mill' the indigo leaves. Instead they shred and winnow them to separate them out from the stems, assisted today by giant electric fans. The stems are made into ordinary farm compost while the leaves are dried. Then, as with the broken up woad balls, the leaf mass is spread deep in sheds to be treated and turned into *sukumo*, the equivalent of 'couched' woad. In Japan the process, which should begin only on a suitably auspicious day in the Buddhist calendar, takes about three months. It is the work of specialists, like the woad refiners of Germany, and a *sukumo* maker used only to pass on his secrets to one of his sons. In the past there were even 'water specialists' who travelled around checking the moisture levels of the *sukumo* beds. Every five days each huge steaming heap, containing up to three tons of leaf material when enough has been added, is sprayed with water and thoroughly turned over, which takes several hours. It is a dirty and arduous job, particularly for a woman who marries into the profession. As with 'woad couching', tools unique to the job are employed, and long experience and fine judgement are crucial to success. Japanese indigo farmers liken the task to nurturing a baby; the thick straw mats that cover the composting leaves are called the bedcovers, and the fermenting chamber the bedroom. If there are problems with fermentation the baby is said to have caught cold. Each time the steaming mass is turned, an offering of rice wine is made to *Aizen Shin*, the god of *Ai* (indigo). If all goes well, after three months the mass will have solidified and darkened, indicating that enzymes and oxygen have done their work.[97] The final substance, *sukumo*, has an indigo content of about 5 per cent (woad's was well below one per cent). This is packed into large rice-straw sacks for distribution. Sometimes it was pounded up and shaped into concentrated balls called *ai-dama* for ease of transport. As with woad, the market price of *sukumo* was set once its dye content had been thoroughly tested.

Until recently nobody knew quite what was going on during the leaf fermentation stage, although a splendid publication of 1675 has diagrams of apparatus designed to collect the ammonia given off by fermenting woad.[98] Organic chemists in several countries are examining the microbiology of the process and some have discovered strains of bacteria that seem to explain what takes place during 'fermentation' (in scientific terms 'the aerobic transformation of plant compounds to indigo by various micro-organisms').[99]

Selling dried indigo dye balls. Iwo, Yorubaland, Nigeria, 1960.

Judging by the account of Jean Barbot and other early travellers in western Africa, the process of indigo manufacture there has changed little over the centuries.[100] Although synthetic indigo now dominates, some dyers in traditional indigo-dyeing regions can still be found who pound up fresh indigo leaves (both *Lonchocarpus cyanescens* and indigofera species are used) in their deep wooden mortars, squashing the leaf paste into cones or into fibrous balls like 'green' woad, which are dried ready for marketing or storing for future use. In Asia dyers of some ethnic groups also preserve indigo leaves this way.[101] Since some fermentation is set off when the leaves are pounded, the dyer's task is simplified.

Oxygenating indigo liquid by whisking with a half gourd after leaves of *Strobilanthes flaccidifolius* had been removed from the barrel and slaked lime added. Fanzhao, Guizhou, south-west China, 1993.

Colonial manufacture

Even in so-called 'primitive' societies many dyers discovered how to extract indigo paste from indigo plants to obtain a more concentrated form of the dye than fresh leaves could offer. The chemical principle was much the same for colonial systems, the main differences between local and 'colonial' methods being scale, control, efficiency and end purpose. Indigo paste produced somewhat haphazardly at a village level contained so much mud, lime and plant material, in addition to water, that its actual indigo dye content could drop to as low as 2 per cent, whereas the best Bengali dyestuff could be over 70 per cent indigo (the rest being indirubin and other organic matter).[102]

Throughout Asia indigo dye paste, which could be stored damp for a year or so and used as needed, was made by fairly simple methods. There was no need to dry it if it was to be used within the area as dyeing is easier using damp paste. Crawfurd in 1820, taking a racist view on the matter that is somewhat embarrassing to read today, declared: 'None of the Asiatic nations are equal to the manufacture of a perfect drug [indigo], fitted for the market of Europe. The Chinese, who can manufacture good sugar, cannot manufacture good indigo, which is the peculiar product of the skill and civilization of Europeans'![103] In fact, of course, the Chinese had been processing indigo for nearly two millennia before this statement was made, and are reputed to have dried the paste when required for long-distance trade.[104] But such was the demand for indigo within China that, far from seeking an export market, the country for centuries imported extra indigo from India, Java and the Philippines.[105] A report of 1898 notes that in China at that time jars of indigo paste were sent from Guangxi province to Shanghai, undercutting the price of Indian indigo.[106] West of Shanghai much indigo was made near Wuhu on the Yangtze river. China clearly managed perfectly well without producing top quality indigo along colonial lines, but, except in the south of the country, she did embrace synthetic indigo (provided in liquid form to suit the dyers)[107] with alacrity when it became available.

Scooping indigo paste, newly made from *Strobilanthes flaccidifolius* leaves, out of the base of a wooden barrel. Zhaoxing, Guizhou, south-west China, 1993.

In indigo-producing provinces in China up until the first decade of the twentieth century every farmer grew indigo and processed it at home in at least twelve large jars or barrels, or in purpose-built cisterns or pits up to two metres deep lined with bricks or concrete for larger scale production.[108] In 1938 Szechwan province still grew indigo on about 5000 acres of land, and some specialist villages in south-west China and neighbouring countries even today continue to manufacture indigo, one Chinese district

Selling damp indigo paste by the roadside. Sapa, northern Vietnam, 1995.

still using huge stone tanks carved out of the mountainside.[109] 'Chinese indigo', indigoferas and *Strobilanthes flaccidifolius* are all treated the same way. Freshly picked leaves are steeped in water, and left until fermentation reaches a critical stage. At this point the liquid is drained off, slaked lime (or in Sumatra and Thailand lime made from shells)[110] is added and the whole is vigorously stirred until enough oxygen has been incorporated for the indigo to form. This precipitate sinks to the base of the container to become a paste which can be scooped out and sold in this state to dyers. On the Japanese island of Okinawa, with its historic links with China and a different climate from mainland Japan, the same system has been used with their own local *Strobilanthes flaccidifolius* (*ryuku-ai*).[111] Here the last professional indigo maker (there were still twenty families after World War Two) makes enough indigo paste, *doro-ai*, to supply all the local dyers.

Producing dye paste this way continued quite satisfactorily everywhere for local use in parallel with the 'colonial' system. In nineteenth-century Java the locals planted for themselves an inferior quick-growing indigofera, squeezing in three or four cuttings between annual rice crops, while the best kind to be processed commercially for export was raised on large European plantations in richer soil and grew more luxuriantly.[112]

Ever since their arrival in India European merchants, especially Dutch and English, were intrigued by the manufacture of indigo and took a keen interest in every aspect, as mentioned in Chapter Two, even trying to get involved themselves in supervising its production. In the early seventeenth century the Indians were already making their best indigo dyestuff in sets of purpose-built cisterns,[113] by steeping the plant and agitating the liquid with 'great staves'. They skimmed off the clear water above the precipitate in stages, until the paste thickened enough to be spread out on cloths and partially dried in the sun.[114] It was then moulded into balls or flat cakes to be baked hard on the sand. Even in these early days European observers realized the value of fertile soil and alkaline water. Finch, for example, described Agra's indigo as being very good 'by reason of the fastnesse [denseness] of the soile and the brackishnesse of the water'.[115] This requirement for alkalinity was recognized by all indigo producers. Paste producers added much lime, while the Europeans later learnt how critical it was to regulate minutely the levels of alkalinity. Finch proceeds to describe the desirable qualities of the best indigo:

Fowre things are required in nill; a pure graine, a violet colour, its glosse in the sunne, and that it be dry and light, so that swimming in the water or burning in the fire it cast forth a pure light violet vapour, leaving a few ashes.[116]

When demand for indigo was escalating rapidly in the eighteenth century, the Europeans set up their own indigo factories in the West Indies and Central America in order to manufacture a consistent product with the highest indigo percentage possible. Instead of using jars or small tanks, they invented an improved version of the Indian system involving much larger cisterns on graduated levels.[117] The French added many refinements to their factories in the Antilles (see p.66). French expertise was later applied in India, above all by the British in Bengal, who, at the end of the eighteenth century, translated into English important French sources in their desire to make Indian indigo as good as that of the Antilles and Guatemala.[118] It is no surprise to learn that when Egypt's ruler, Muhammad Ali, monopolized the Egyptian indigo industry in 1824 he brought over forty Armenians from Bengal to establish costly colonial-type factories.[119]

Supervision was indeed the key to industrial production – both for controlling a reluctant labour-force and for overseeing every stage of the process. The indigo crop had to survive the vagaries of climate, diseases such as wilt, and insect damage[120] (although Diderot mentions that the chewing of leaves by caterpillars could be beneficial),[121] and then had to be meticulously processed to ensure a successful outcome. In the words of Monnereau, 'in a bad season the most skilful indigo-maker is likely to be baffled'.[122] Unless leaves were being properly dried for later use, it was vital for harvested indigo branches to reach the clean cisterns without delay as quality deteriorated if they were left hanging about. The largest, upper cistern was packed with branches, and water was added. After an average of twelve hours fermentation, enzymes would transform indican into indoxyl, as indicated by a dramatic frothing up of bubbles which changed colour from white to purplish. The critical moment had arrived to drain off the olive-yellow liquid into a lower tank, leaving the rotting leaves behind in the upper cistern.

Then came the really unpleasant job, which needed to start immediately. Groups of near-naked men and women had to walk up and down waist deep in the slimy liquid for several hours, beating it with implements such as wooden paddles, or even with their bare hands, to stimulate the oxidation of the indoxyl. The relevant plate in Diderot's *Encyclopédie* aptly labels one tank *le diablotin*, 'devil's tank', as the terrible fumes, according

Indians beating indigo liquid by hand with wooden paddles at a European indigo factory in Indian Bihar, 1898.

to the author, 'killed many workers'.[123] Local populations were understandably reluctant to undertake a job said to cause, if not death itself, at least cancer, impotence, headaches and temporary lameness.[124] (In India only certain castes would do this job.)[125] Some people, notably the American Indians, were thought to be more susceptible to illness than others. The consensus was that African slave labour was not only more expendable (easily replaced) but also physically more suited to the task. Those who had already survived the appalling conditions under which they were shipped from Africa must certainly have been tough.

Detecting the exact moment at which to stop oxidation demanded great experience. In Spanish America respected professionals called *punteros*, 'point watchers', were employed solely to detect subtle changes at this stage of the manufacturing process.[126] In India the controller responsible presided over this stage as at a 'mystical ceremony';[127] and Diderot describes the 'artist' with his special polished silver testing cup.[128] If beating ended too soon or went on for too long, the quality and quantity of dye was adversely affected. One famous French source in Mauritius dwells at length on this delicate stage of production, saying that every single batch of plant material, even the same species from different parts of the same field, had its own point of optimum fermentation.[129] The writer also considered that the size and shape of the cistern made a difference, and that adding an alkaline substance helped to achieve 'simultaneous fermentation' throughout it. He advocated wood ash rather than lime, noting that even ash from different ages of wood had different effects, and mentioned additions such as sugar and soot, and tests which included 'swimming a fresh egg' in the liquid. Once the critical decision was made and beating ceased, the liquid, which would have changed from green to dark blue, was left undisturbed to allow the *fecula* (indigo precipitate) to settle. Then the water was drawn off in stages, leaving behind the clay-like blue paste. This was removed, filtered and strained through coarse cloth, and cut into blocks when almost dry.

The degree of sophistication in the process varied from country to country, and of course improved over time as scientists and engineers devised ever more efficient methods to extract indigo and control its quality. The most significant of the advances were ingenious inventions to aerate the liquid mechanically. Sometimes horses or mules drove the paddles, or in the larger works water wheels were used. Later came steam power. In nineteenth-century Bengal and Bihar indigo plants were processed beside the Ganges either using man-power in rows of up to twenty large tanks, with a matching series of shallow tanks, or mechanically with a beating wheel in one very long tank. Larger factories included pumps to convey the indigo sludge to cauldrons heated by steam-driven boilers and engines; here the paste was boiled to prevent further fermentation. After filtration the buttery mass was gradually compressed in special screw presses to a depth of about 9 centimetres before being cut into soap-sized cakes, stamped with the date and factory mark and sent off to the drying house, a building carefully situated and designed to prevent the paste drying too quickly and cracking. In the humid season drying took three months, then mould was brushed off and the cakes were packed into mango-wood chests.[130] This method produced approximately 25 kilos of indigo per fermentation tank. The whole procedure, from the plants arriving by bullock cart at the steeping tanks to blocks of indigo emerging from the factory, can be viewed in miniature at the Royal Botanic Gardens at Kew, London, where there is a splendid model of an indigo factory, one of three made specially for the British Colonial and Indian exhibition in 1886.

Ever since Pliny's warnings about indigo adulterated with clay or pigeon's droppings (see p.23), manufacturers continued to be tempted to add bulk and weight to their product, using substances ranging from rubber and brick dust to Prussian Blue and tiny particles of blue cloth.[131] (Similarly, woad farmers would add bulk to woad balls with thistles and other weeds.)[132] Merchants' handbooks constantly advise on methods for testing commercial indigo.[133] Best varieties had a fine texture and purplish hue, adhered to the tongue when licked, appeared coppery when scratched, 'swam' in water and when burnt 'flew like dust' and gave off a violet-coloured smoke.[134] There were trade descriptions for all the different varieties from each country, but as a rough general guide indigo was categorized, in order of merit, into 'sandy, ribboned, spotted, burnt, large squared, half-broken, coarse granulated, cold, and fig'. The worst kind (Manila indigo) would often be used for laundry blueing.[135]

Quality control may have affected merchants' pockets, but was of equal concern for the dyer trying to maintain consistent results and having to pay large sums for his raw materials. Nineteenth-century chemists employed by large textile-dyeing and printing firms devised ever more ingenious tests to establish indigo's varying qualities more scientifically.[136] The matter was complicated by the fact that dyers, as mentioned, often preferred indigo which contained certain so-called 'impurities' which they actually found beneficial (see p.84).

Indigo cultivation today

Small-scale production of natural indigo has survived in various places already mentioned. There are several instances of the industry being revived, sometimes in response to a demand in the niche market.

One German enthusiast runs a project that uses Turkish woad for dyeing modern carpets.[137] He is also encouraging farmers in El Salvador to produce indigo in tanks constructed in Spanish colonial days, but using more ecologically sound techniques. The Wissa Wassif weaving project in Egypt and 'Renaissance Dyeing' in Wales[138] are two other organizations who prefer to use natural indigo whenever possible in their quest for historical authenticity. Such projects are unusual because elsewhere nearly all those claiming to use 'natural dyes' in fact always use synthetic indigo for blues as it provides such a tempting shortcut. In Korea one young farmer who is reviving traditional methods of natural indigo production has, like his Japanese counterparts, received government recognition.[139] More encouraging still, in Guinea scientists have developed a more efficient way of producing cheap natural indigo in powdered form.[140]

In western Europe too the future of woad looks promising. At *Bleu de Lectoure* in French Gascony indigo dye pigment is processed from fresh woad leaves by Henri and Denise Lambert, who use it both for dyeing and to manufacture art materials and household paints. Other European projects are in the pipeline thanks to selective breeding programmes and innovative biotechnological research being carried out in England.[141] These should provide farmers with an alternative source of income, and global climate changes could well have an impact since woad leaves yield far more indigo when harvested after several days of sunshine. This is good news for southern England, for example, which now has a sunnier climate more akin to the woad-growing regions of southern France in the Middle Ages. However, a word of caution for beekeepers: if medieval sources are to be believed, beehives should be sited well away from fields of woad as the plant was blamed for giving bees the 'flix', or 'flux' (i.e. loose stomachs).[142] As for the possible future of indigo in an entirely new form thanks to advances in genetic manipulation, see the final chapter of this book.

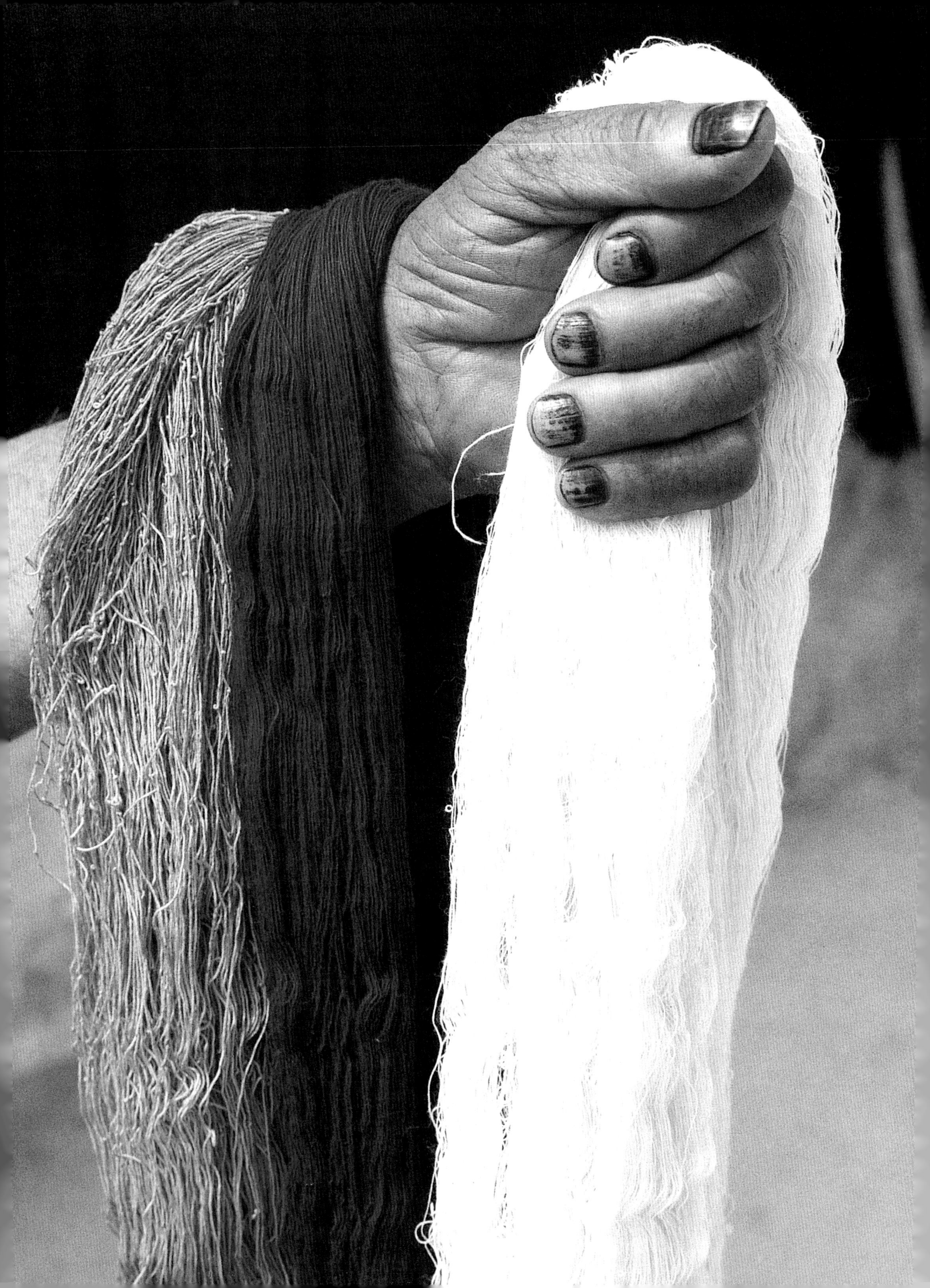

Dyeing is an art; the moment science dominates it,
it is an art no longer, and the craftsman must go back to the
time before science touched it, and begin all over again.
(ETHEL MAIRET, 1916)

Blue Nails

Indigo Dyeing Worldwide

There is something very mysterious about a dye that only reveals its colour *after* yarn or cloth emerges from the dye pot. Now that chemists can explain the scientific rationale of indigo dyeing, an element of its magic has gone. Yet even those with a chemical background cannot fail to be enchanted when they first see the slow transformation of yellow into blue take place as newly dyed cloth is removed from an indigo dye vat and oxygen turns sorcerer.

So what makes indigo stand aloof from other dyes? The answer lies in its idiosyncratic chemistry. Some natural dyestuffs are 'substantive', or 'direct', meaning that with heat they will fix directly to a fibre with which they have an affinity. However, most natural dyestuffs belong to a large group known as 'adjective', which require an intermediate chemical substance, called a mordant, to make them fast. Pliny marvelled at their use by the Egyptians in the first century AD; indeed, archaeological finds indicate their use some 2000 years earlier in India. A mordant is a metal salt, or the organic equivalent, which links permanently with the molecular structure of a fibre when dissolved in water. The dye itself bonds with the mordant during the dyeing process. In this group are found most major natural dyes, i.e. insect and vegetable reds, most plant yellows and certain mineral substances. Colours produced by mordant dyeing vary greatly depending upon which mordants are used, the basic ones being alum, iron and, later, chrome, although there were many other organic ones in the past.

Indigo dyer's hand holding undyed yarn, and yarn dyed pale and dark blue in the indigo vat. Sayeda, near Sousse, Tunisia, 1995.

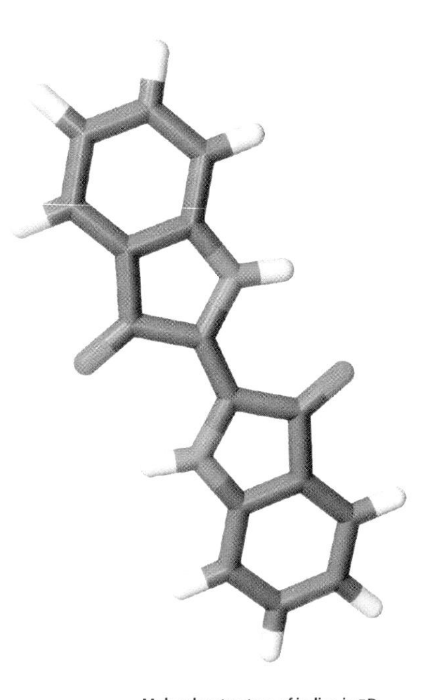

Molecular structure of indigo in 3D.

Indigo is completely different from all the other natural dyes, apart from its close cousin, shellfish purple. It needs no mordant to make it fast to light and washing, it is insoluble,[1] and is deposited on the fibres as microscopic particles without needing to form a thorough chemical bond with them. This makes it uniquely suitable for dyeing any type of fibre, but it needs special handling. The first stage of the dyeing process is to dissolve and reduce the insoluble dyestuff in a warm (about 50°C) alkaline solution in a 'vat' (named after the deep wooden barrels used by medieval woad dyers). This adds hydrogen and temporarily transforms indigo into the leuco-derivative sometimes known as 'indigo white', although the dye liquid in fact looks yellowish-green. In the vat this is deposited on the surface of the immersed fibres so that when they emerge they too are yellowish-green in colour and only turn blue after several minutes exposure to the air, when oxygen converts the dye back to its permanent blue form again. This unusual series of chemical reactions has given rise to much of the mystery associated with indigo dyeing. The underlying chemical principles never change despite very wide variations in the making of indigo dye vats. These depend on local resources and traditions, the dye source itself, the form of the dyestuff (fermented leaf material, extracted dyestuff or synthetic indigo) and the fibres to be dyed. Small mysteries that still remain unsolved relate to micro-bacteriological reactions in the organic dye vat, but in the days when dyeing was a purely empirical process it is small wonder that the stages of chemical transformation seemed so baffling. When dyeing was becoming more scientific, indigo was still considered the most difficult of all dyes to use. The renowned craftsman William Morris found it so challenging that he wrote to a friend: 'It would be a week's talk to tell you all the anxieties and possibilities connected with this indigo subject ...'[2] However, just like all his predecessors, he found that indigo dyeing repaid the trouble taken as the end result is inimitable.

There are a few exceptions to the concentrated 'vatting' of indigo dyestuff which do not produce such strong or permanent blues. Apart from the use of fresh leaves in the vat, a direct method invented in 1740 involved mixing ground indigo with sulphuric/disulphonic acid or oil of vitriol to make 'sulphate of indigo', 'extract of indigo' or 'indigo carmine'. The resulting bright turquoisy blues were known by wool and silk dyers as 'Saxon' or 'chemic' blue.[3] Yarns dyed in this way can be seen in some oriental carpets, most characteristically those made in Turkey during the second half of the nineteenth century,[4] and also in late eighteenth-century Kashmir shawls. In the Solomon Islands an unorthodox method has been recorded whereby juice created by masticating indigo leaves and lime was spat directly onto bark cloth and worked in with the fingers.[5] The Amazonian Indians, too, who paint bark cloth with local indigos, discovered the value of saliva in producing a kind of indigo by chewing on a certain fruit.[6]

Cloth in an indigo vat made from fresh *Polygonum tinctorium* leaves in Devon, 1994. The surface has a copperish film while the liquid beneath is yellowish-green with indigo in its reduced state.

Indigo is the most versatile of all natural dyes. The majority, including animal dyes like insect reds, have an affinity with the proteinaceous fibres of wool and silk which are damaged by alkaline substances. The alkaline medium of the 'exotic' indigo vat was particularly well suited to cellulosic fibres like cotton and the bast fibres (used since ancient times) of flax, hemp and ramie. None of these fibres easily absorb most natural dyes unless first subjected to laborious processing, as was necessary to make 'Turkey red' from madder. Indigo, though, will dye all such fibres, hence its use as the earliest dye found on archaeological textiles of linen. Indigo also dyes animal fibres of silk and wool beautifully but they need a less alkaline vat, particularly in the case of wool, which prefers a pH of around eight. For this reason the gentler woad vat of the past was particularly suitable for dyeing wool. For centuries, moreover, indigo provided a really dark colour that may even have strengthened the fibres and certainly did not damage them as did the browns and blacks made from iron and tannin.

Whatever the fibres used, in the case of indigo, again unlike other dyestuffs, the subtlest and deepest shades can only be produced by multiple immersions in the vat to build up layers of colour through stages of dyeing and oxidation, leaving the actual core of the fibre undyed. There are no short cuts, as dyers with a true feel for indigo soon come to appreciate. (Chemically speaking, the sheets of flat hydrogen-bonded molecules are layered. The bonds are strong, hence indigo's fastness to light, but the layers slide apart relatively easily, hence its susceptibility to rubbing. Soaping, polishing or gumming dyed cloth may have helped to minimize rub-off.) Once dyed a deep blue, indigo is so colourfast that it can last for centuries, or even millennia, as is evident on archaeological textiles. The deep indigo-blue fields of oriental carpets such as the famous Persian sixteenth-century Ardabil carpet in London's Victoria and Albert Museum are ample proof of the dye's durability. When indigo-dyed fibres do gradually fade, especially after extensive washing or rubbing, they retain a special quality, thanks to the layering of the dye. This means that unlike most natural dyes indigo, however old and faded, never loses its hue. This characteristic is frequently exploited deliberately,

notably to produce the 'stone-washed' look of ever-fashionable blue denim.

Although chemists know exactly what is going on in a modern indigo dye vat, successful dyeing still depends upon finely controlled balancing of all the variables: i.e. pre-treatment of fibres, concentration of dyestuff, water temperature, alkalinity, amount of reduction, rapidity of oxidation, length of immersion time and number of immersions, methods of circulating yarn in the dye bath and efficiency of chemical replenishment.[7] Even today when industrial dyers have at their disposal stable synthetic indigo (in the form of grains, paste or pre-reduced indigo solution) and modern chemicals, dye manufacturers and the dyeing industry are still seeking ways to standardize the process. Recent BASF literature on indigo-dyeing yarn for denim states:

Irregular dyeing results – particularly variations within dyeing lots – are a problem in denim production. As a rule, the finished denim material is classified into groups according to the dyeing results and the appearance of the wash-down. Maintaining the dye-bath parameters constant reduces the number of these groups and also the proportion of 'seconds'.

To regulate these 'parameters' the recommended pH is 11.5, as below that the shade goes duller and greener. On the other hand, too little reducing agent causes a reddish dullness and loss of penetration.

If today's dyers of denim still have to take such care, no wonder people over the centuries held indigo dyeing in such awe! By whatever means the dye potential of indigo was discovered in any society, all the variables had to be monitored and balanced empirically without the aid of litmus paper, thermometer or modern chemicals. We can only admire the ingenuity of a dyer who had to rely entirely on taste, appearance, smell, 'feel' and years of experience, to manipulate intrinsically unpredictable organic ingredients. Many of these sound quirky or superfluous yet prove on closer inspection to have sound chemical functions. The reluctance of indigo dyers the world over to divulge their secrets is easy to understand, as is the existence of strange beliefs and myths that came to be associated with indigo. The more one looks into the indigo dye vat, the less one is inclined to laugh at the superstitions of people in so-called 'underdeveloped' societies.

Keeping account of the number of immersions in the dye vat by chalking them on a blackboard. Workshop of Josef Koó, Austria's last 'blue-printer', Steinberg, 1995.

Bark cloth from the Solomon Islands decorated with indigo juice spat onto the cloth and painted on with the fingers.

The fermentation vat

Throughout the world, and whatever the plant source, the fermentation vat was the only option available to dyers before the sixteenth century, when experiments with inorganic reducing agents got under way to speed up the process. In some parts of the world the fermentation method has continued largely unaltered down the centuries and can still be found today.

Like leavening bread or making beer, a procedure based on bacterial fermentation is 'alive' and cannot be hurried as the microbes need time to do their work. In the case of indigo this involves converting the dye into a soluble, reduced state.[8] Nancy Stanfield summed up her recollections of the elaborate preparations for indigo dyeing by Yoruba women in the 1950s as 'a time filled with superstitions and a lot of stirring, sniffing and sitting around waiting'![9]

Nourishing the dye vat with organic substances is quite a challenge, as innumerable variables can affect the process, not least the local climate. In a rural setting indigo dyeing was, like weaving or house-repairing, a seasonal activity, usually undertaken when agricultural activity was slack or, if fresh indigo was being used, in the rainy season when plants were at their best.[10] In a large-scale urban context the industry would be a continuous activity in large workshops employing many people. Either way, vessels for woad and indigo dyeing, unlike those for mordant dyeing, did not require direct heating, but needed to be large and good insulators if dyeing was to be even. They were invariably made of wood or clay, as opposed to metal which could taint the contents. A dyer often had several dye vats on the go at once. Although dark shades can only be built up over several dippings, in order to dye cloth exceptionally dark for an important textile such as a chief's robe, a freshly made dye pot was used for each dipping. Dye vats could be kept 'alive' for days, months or even years, depending both on local customs and on methods used.

The easiest fermentation vats to make are those using pre-fermented, composted plant material. There is no hard lump of insoluble dye pigment to dissolve, and bacteriological reactions have already been activated during the 'couching' stage. This is why the medieval woad vat, and those still surviving in West Africa and Japan, are comparatively quick and easy to prepare. Dye can be ready to use in a matter of days, whereas vats made from hard indigo dye pigment can take up to two weeks to prepare.

In the Middle Ages only 'couched' woad and a wood-ash lye were absolutely necessary to make an effective vat for dyeing wool, according to the oldest complete recipe from an Italian *Arte della lana* of 1418,[11] recently transcribed by Dominique Cardon.[12] This recipe states that the vat was set with 400 pounds of woad to twenty barrels of water. Over a

Yoruba dyers preparing lye by filtering water through broken up wood ash balls placed on a mesh of twigs and leaves in the upper clay pots. The alkaline liquid was collected in bowls inserted though holes in the lower pots. Nigeria, 1960s.

hundred pounds of oak-wood ash and more hot water were added stage by stage, with much 'raking' and checking for visible signs such as a 'colour of gone-off sauce'. 'A connoisseur would know from far away, by his nose, if the colour was on the verge of going off,' in which case hot madder dye or bran could be added to rescue it. On the second day the dye would be 'ripe' for use. Other similar medieval recipes studied by Cardon took a day or two longer to make, and some included wine lees.[13] The woad vat became exhausted after a few days' use, whereas post-medieval vats which included imported indigo could be periodically revived and kept on the go for long periods.[14] (Dyers often retained some old dye vat liquor to 'start off' a new one.) From the thirteenth century, when dyers in southern Europe began to make separate vats of higher alkalinity for dyeing linens, cottons and silks, they used imported indigo ground into a powder and added to what Cardon terms a 'madder soup' made from bran, madder and calcinated lees.[15]

As with the medieval woad vat, traditional vats of West Africa were based on dried balls of crushed leaf (up to a hundred per vat), made either from *Lonchocarpus cyanescens* or from indigoferas, along with paste from the base of an old vat and ash lye commonly made from the ash of burnt coconut fibre, millet stalks or indigo branches themselves. A fifteenth-century Italian dye manual dwells at length on the differing qualities of ash lyes;[16] similarly African dyers would know from experience which ashes were best (i.e. highest in potassium carbonate). If the Frenchman Jean Barbot's account of preparing indigo in Senegal in the 1670s is accurate, at the time he was writing the dried indigo balls (obviously

Dyer in Bakel oasis, Senegal, with a large cone-shaped lump of dried indigo made by pounding indigofera leaves in a wooden mortar, 1997.

of an indigofera species as he compares them to 'wall-rue') were processed with indigo-plant ash by an all-in-one method. He describes how 'clear spring water' was gradually filtered through a layer of pounded-up indigo balls covered in ashes. The resulting liquid was left in the sun for ten days which, he writes, 'thickens the liquor in it, like cream, the top whereof they take off gently, and dye with it.'[17] Later the ash lye was prepared separately. In Nigeria the dyers of Yorubaland turned its preparation in sets of ceramic pots into a high art.[18] In other places today lye is made by filtering ash water through holes in a bucket lined with straw, and the lye when dried doubles as cooking soda. Sometimes African dyers add such extra ingredients as the crushed root of *Morinda geminata* or pounded kola nuts, probably partly to enrich the colour.[19] As the whole process takes up to two weeks, such vats, prepared just as Mungo Park described in 1805,[20] are getting less and less common today. In Oshogbo, one of Yorubaland's formerly renowned indigo towns (others being Abeokuta, where 80 per cent of the population was employed producing indigo-dyed textiles in the past,[21] and Ibadan), only one dyeing family remains, thanks to the enthusiastic encouragement of Nike Olaniyi Davies. To the north Kano and the Bornu region used to attract buyers from all over the Sahara. The dramatic decline is evident throughout West Africa, although traditional dye vats can still be found in specific areas in all its states.[22]

In Japan, too, *sukumo* or *ai-dama* (i.e. composted leaves of *Polygonum tinctorium*), was sometimes just mixed with ash, but fermentation would then take up to thirty days in cold weather.[23] To speed up the process Japanese dyers will add rice bran, rice wine, honey or lime, and the sunken vats (over 200 litres capacity) are often grouped in sets of four so that a fire can be lit between to gently warm the dye.[24] The Japanese dyer, like the medieval woad dyer, calls the scum that floats on the top of a healthy dye vat the 'flowers'. These flowers can be scooped off and used as paint pigment, or pushed to one side before dyeing begins.

In parts of the world where dye pigment or paste, or even fresh plant material, was used, a complicated witches' brew of ingredients was often concocted. For example, one researcher found rural indigo dyers in Ecuador in the 1980s using no less than twelve ingredients in their dye vats.[25] In some places dye vat ingredients have never been fully recorded.

When listed together, the substances found to be effective reducing and fermenting agents by different societies worldwide sound more like the ingredients for an extremely elaborate festive cake, since many of them are 'sweet'.[26] They have included dates, grape and palm sugar, molasses, yeast, wine and rice spirit lees, local liquors, beer, rhubarb juice, figs, mulberry fruit, papaya, pineapple, ginger, honey, jaggery, henna leaves, wheat bran, flour, cooked glutinous rice and tapioca, madder, *Cassia tora* seeds, sesame oil, green bananas, sisal leaves, powdered betel nut, tamarind juice and, less appetizingly, putrefying meat. Sometimes extra meat or fruit, although ostensibly added to propitiate threatening spirits, would in fact have genuinely aided fermentation. Meanwhile alkalinity was provided by natural soda, lime, wood ash, soapwort, alkaline river mud, chicken droppings, animal turds and, last but definitely not least, stale urine. Slaked lime was the most effective at dissolving indigo pigment. Alkalinity is essential to neutralize excesses of lactic acid caused by fermentation. As old dye manuals put it, the vat must not be allowed to 'bolt' or 'crack'. Some of the above substances would have added additional colour in their own right to enrich or darken the indigo blue. These include madder root, cutch (catechu), camwood, woad (in indigo vats), the *sacatinta* plant (*Jacobinia Mohintli* Benth. et Hook) in Central and Southern America,[27] *Cassia tora* seeds and *Morinda tinctoria* wood in India, wood chips in south-west China, peach and other barks in Japan and Indonesia, and betel nut and turmeric on Savu island. The precise function of some of the more bizarre ingredients – amongst which the juice obtained from boiling up a nest of red ants should perhaps take pride of place[28] – occasionally baffles organic chemists, but the function of the majority can be understood. Seemingly irrational superstitions concerning dyeing usually reveal an instinctive feel for basic chemistry. To give just one example, on the Indonesian island of Flores any contact with sour fruit (i.e. acidity) is strictly forbidden near the dye vats, and dirty hands must be cleansed with ashes.[29]

Though surviving elsewhere, too, the biggest concentration of fermentation vats today is in Southeast Asia, especially among ethnic minorities of an area taking in south-west China, Thailand, Burma and Vietnam, and on islands of eastern Indonesia.[30]

The transition period when dyers were edging away from fermentation to inorganic vats has often been gradual and haphazard as new chemicals appear

Dyeing ikat-tied yarns in indigo. Ba Saang Sa, village weavers' project, near Nong Khai, north-east Thailand, 1991.

Set of four deep dye vats with 'flowers' floating on top of the dye liquid. In cold weather a fire is lit between them. Workshop of Hiroyuki Shindo, near Kyoto, Japan, 1993.

Indigo dyer at Imbabura, Ecuador, 1980s. Sometimes up to twelve different ingredients were used in the dye vats.

on the market without precise instructions about their use. At first a dyer may make a traditional vat but throw in a touch of synthetic indigo at the end to strengthen the colour, rather as the medieval European woad dyer first added Indian indigo to his woad vats. Dyers in more urbanized areas began between the two World Wars to get hold of modern chemicals that would speed up the whole process of dyeing, especially for the sought-after dark colours.[31] Today some dyers who have switched to modern methods will occasionally throw in traditional fermentation ingredients, seemingly for old times' sake. Or they have moved largely to cheaper black dyes but remain reluctant to fully relinquish indigo. They will therefore sometimes dye a cloth in black but give it a final dipping in indigo. In such places, as far apart as Ecuador and Yemen, cloth or yarn dyed only with indigo is still considered the best, black overdyed with indigo being second quality, and pure sulphur black definitely the poor relation.[32] But few members of the present generation wish to be indigo dyers like their parents. Today in Sierra Leone the West African word *gara*, for indigo (once virtually the only dye) is now applied to cloth hand-dyed in any synthetic colour.[33]

Women planting garlic in the Hadramaut valley of eastern Yemen, 1985. Many of them were wearing 'black indigo' dresses, dyed in cheap black dyes but given a final dipping in a traditional indigo vat.

Urine vats

As a dye vat ingredient, stale urine (which produces ammonia), whether human or animal, deserves a special mention. If only someone could invent a method of removing its smell it might even be widely used today, as it is most effective and always freely available! It provides an all-in-one reducing agent and mild alkalinity especially suitable for wool fibres.

The urine vat has a long history. When in the 1930s the block-printers Barron and Larcher had a pail on hand for amenable male visitors,[34] they were carrying on a habit dating back to Classical times. Dyers of Roman Pompeii placed a pot outside their dye shops to secure this ingredient from willing passers-by. Greek papyri of the third century AD from Upper Egypt (the *Holmiensis*, or Stockholm Papyrus, and the *Leidensis*) recommend urine both from animals and from boys before puberty. Although different types of urine may have contained different strengths of ammonia, that of young boys probably had more to do with virginity, which was in antiquity considered to bestow special qualities.[35] A recipe in the *Holmiensis* papyrus, found in 1928 by archaeologists in the tomb of an alchemist-priest in the necropolis at Thebes,[36] gives instructions for a fairly standard urine woad bath, with an addition of soapwort to improve dye absorption and a final dipping in the reddish-violet lichen dye, orchil (*Rocella tinctoria* D.C.), to enrich the colour.[37] The text, with the occasional obscurity, first describes the harvesting and preparation of woad. Batches of 'one talent' (about 25 kilos) of dried woad, termed *kohle*, were apparently immersed in urine in a 600-litre capacity tank where they were soaked and trampled on in sunshine over three days. This mixture was divided into three and made into the following dye bath:

Stir up the soaked woad properly, put it in a pot [vat] and light a fire under it. The way to tell whether the woad is cooked is the following. When it simmers, stir it carefully and not too hard, so that the woad doesn't sink down and ruin the vat. If the woad parts in the middle, it has been heated sufficiently. Remove the fire, but continue stirring as before. Cool the underside of the vat by splashing cold water on it. Then take half a choimix *[½ kilo] of soapwort and put it in the vat. Place the plant stems around the rim of the vat, cover it with mats and make a moderate fire under it so it doesn't get too hot or become too cold. Leave it for three days. Then boil up urine with soapwort, skim off the scum and put in the washed wool. Then rinse it carefully, squeeze it out and put it in the dye bath. When it looks ready, take the wool out, cover the dye bath again and make a fire under it as before. Put into the liquid 2* minen *[1 kilo] of orchil, boil it up and skim. Now add the dyed wool. Rinse it in salt water and let it cool down. Do the blue dyeing twice a day, morning and evening, as long as the dye bath is usable.*

No wonder the Greeks complained that 'the hands of the dyer reek like rotting fish'.[38] Dyers, however, despite the objectionable smell, have been prepared to hold their noses for the sake of the end result. Many medieval European woad recipes specify urine as the main ingredient, noting that the vat should be scalding but not actually boil.[39] Its suitability for wool led to its continuing use for domestic dyeing in northern Europe in the first half of the twentieth century. In the Highlands and Islands of Scotland housewives still kept a urine-based woad or indigo dye pot permanently on the go, warming it among the peat ashes in the fireplace, and using it to dye woollen yarn for Harris tweeds and fishermens' jerseys.[40] Even today a few dedicated craft dyers still swear by it. An unusual recipe was recorded in Ecuador in the late 1980s which calls for 7 litres of ash lye, 4.5 litres of urine and the juice of two crushed sisal leaves, the whole concoction taking two weeks to 'mature'.[41] Claims have been made for the superior efficacy of urine from male youths (linked perhaps with their alcohol consumption).

Last indigo dyer of Ibri, Oman, standing by his newly dyed cloth in the old suq, 1985. He was using natural indigo cultivated nearby.

This may be a suitable point to mention the role of animal turds in the organic dye vat. In the seventeenth century de Thevenot praised the effective use of a solution of dog's turds by the indigo dyers of Aleppo, who, he says, 'make a most excellent blew dye'. He describes the preparation of the vat over many days, based on indigo, pomegranate peel, date juice and the turd solution. He writes: 'They put no urine to it, using dogs-turd instead of it, which they say makes the indigo to stick better to things that are dyed.'[42] Recipes in the sixteenth-century Italian *Plictho* dye manual recommend additions of chicken droppings to the woad vat, and horse manure, 'warm in putrefaction', for tanning leather.[43] In Britain in the 1930s dogs' turds were still commonly used by tanners, as some who worked in the business can even recall today.[44]

The sulky vat

Dyers the world over have attributed a spiritual dimension to dyeing processes,[45] but especially those using the capricious indigo. Being 'alive', it is no surprise to find the organic indigo vat described in anthropomorphic terms. As it can be temperamental and difficult to handle, and has mysterious phases of fertility, the indigo vat has usually been considered female. Early American and European dye manuals would make remarks concerning the indigo vat such as 'If she be a little too hard, it will suffice to let her remain quiet four, five or six hours', and 'Observe not to hurry her'.[46] Similarly, a Frenchman earlier this century described the need to calm an 'angry' vat in Tunisia.[47] In Java a dye vat will 'sulk' when a husband and wife quarrel, and has to be won round with offerings.[48]

Indigo dyers worldwide have looked around for someone, or something, on which to pin the blame when a fermentation vat inexplicably ran into trouble, and what more time-honoured custom is there than to blame the wife or mother! In Ireland, however, the usual sex discrimination was reversed, for female dyers considered a male presence unlucky while they were dyeing with woad,[49] and in the Himalayas any stranger coming too near the vat got the blame.[50] However, in many cultures, notably but by no means only those of Indonesia, it is the powerful fertility of certain women that has been seen as directly conflicting with the fertility of the dye vat, which is equated with a womb that harbours life.[51] For this reason the delicate process of preparing the dye, considered akin to conceiving and bearing a child, has often been reserved for women beyond the age of child-bearing, as those able to bear children could cause the inexplicable 'death' of a dye vat. In Bhutan it was thought that if a pregnant woman dyed yarn in an indigo vat the baby in her womb would 'steal' the colour from the dye pot – a kind of inverse baby-snatching. On the Indonesian

Imprint, in indigo dye, of the hand of the dyer on the wall above the dye vats to ward off evil spirits. Sayeda, near Sousse, Tunisia, 1995.

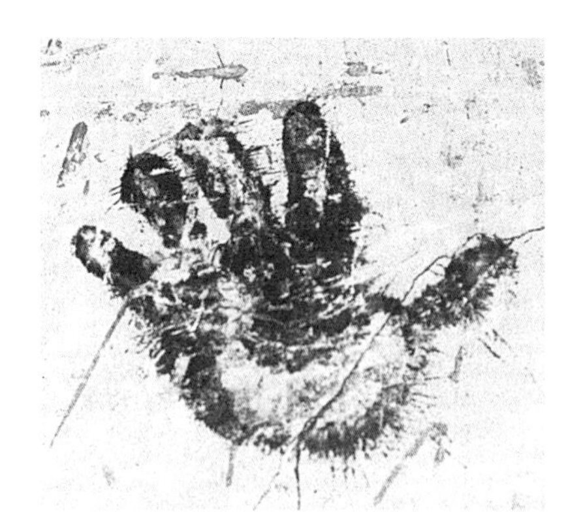

Indigo dye pots, one covered with bamboo leaves to protect it. While covering the pot, the Toba Batak dyer muttered a special phrase to 'blind' the evil spirits. Northern Sumatra, 1980.

island of Sumba, where terms for indigo dyeing and pregnancy are shared, midwife and indigo dyer both observed similar rituals, such as the sacrifice of a chicken and the examination of its entrails, on the occasion of childbirth and during dye preparations.[52] On the neighbouring island of Roti, where an amulet of chicken feathers was once suspended over the dye pot to prevent evil spirits from dipping hands or breasts into it and thereby ruining the dye, Christian dyers now believe that the sign of the cross will protect their dye.[53] The Toba Batak of northern Sumatra, like dyers of Thailand, would, if there was a death in the village, rush to cover their dye pots. This would prevent the spirit of the dead from invading the dye vat, and protect the vat's own spirit, thereby saving the dye itself.[54] Indigo dyers in Puebla in southern Mexico told one researcher that they added touches of lime to the vat just 'to keep away the spirits of the dead ones'.[55]

Offerings to the spirits that reside in the dye vat have often been made while the dye was being prepared, or the dye pot has been 'fed' to please them. Many substances used to induce fermentation were, as we have seen, also normal foodstuffs. In some places the dye vat was 'offered' the same food as that exchanged at pregnancy ceremonies.[56] On Indonesian Savu a black chicken, especially raised for the purpose, was sacrificed before dyeing procedures began,[57] and in western Sumba ikats dyed with natural indigo were far more expensive than those dyed with synthetic indigo, because the former entailed the sacrifice of a piglet.[58] With or without spirits, the vat's appetite seemed human. A medieval woad European recipe states: 'The woad-vat is like man, that wants to eat and drink, otherwise he dies.'[59]

Japanese dyers have a close relationship with indigo, and in their language the spoken word *ai* means both 'love' and 'indigo'. At the beginning of each year a Japanese indigo dyer makes a representation of the Buddhist guardian deity of dyers in the form of a miniature paper kimono (the kimono being a potent symbol of national identity) partly dipped in indigo (see p.i).[60] This is placed in the workshop shrine to preside over dyeing operations. The Japanese indigo artist Hiroyuki Shindo explains: 'Indigo dyers trust in the god of indigo, Aizen Shin, who is enshrined in our workshops and to whom we pray for luck before starting our work.'[61]

In some societies the indigo gods were so feared that indigo dyeing was akin to witchcraft. According to Rumphius, writing in the mid-eighteenth century, the Chinese were ambivalent about the types of gods who guarded the dye vat, difficulties with dyeing being attributed to 'an evil genius or the evil eye'.[62] Writing about Flores (where the making of lime from coral was also spiritually dangerous) one researcher has described the rites that took place before communal dyeing. One woman, noted for her magic, served as leader:

By the use of special phrases, she would chase away evil spirits before the pots were set in place. Under each pot were placed seven threads and seven pieces of 'iron', so that if someone had died, his or her spirit would not be able to touch the dye mixture in the pots. People say the threads will not turn black if the ghost of a person who has just died is hovering around. If anyone's threads were not taking the dye properly, then the leader would use further magic to chase away the spirits of the dead.[63]

In Senegal, before dyeing began in St Louis, the local marabout would bestow a charm to be buried near the vat while prayers were chanted and special plants burnt as offerings to appease the djinns, or spirits.[64] On the other hand, the Yoruba women of Nigeria, for whom every activity has an underlying religious significance, would cease indigo dyeing every sixteenth day in order to worship Iya Mapo, an important goddess who protects female trades.[65] If something still went wrong, blame for failure could always be pinned on chickens passing too near the vat.[66] (Why do chickens so often seem to be accused – could this be connected with the strength of fresh chicken droppings?)

Other elaborate excuses to explain a failed vat include a curious conviction, common to dyers of southern India[67] and the High Atlas mountains of Morocco, that telling lies would charm the potency back into a failed vat. Thus a female Berber dyer in Morocco would deliberately spread malicious falsehoods (i.e. blackness) around the neighbourhood in order to revive her vat by outdoing the malevolent spirit that was thwarting her efforts (a kind of homoeopathic lie!). So widespread was this habit that anyone hearing a malevolent rumour would declare 'That is another lie of the dyers' or 'Ah – there's a failed vat'.[68]

Development of the inorganic and modern dye vats

The first European attempts to simplify the laborious fermentation process were linked with the increased imports of Indian indigo in the sixteenth century. An Indian method of dyeing using arsenic sulphide in the vats was tried out by European dyers, who called this the 'orpiment vat'. But the difficulties they experienced in handling this corrosive ingredient provided ammunition for the smear campaign launched by those who profited from the woad industry and were therefore intent on keeping indigo out at all costs.

It was, though, thanks to woad dyers themselves that indigo became more popular, as they began, mainly from the sixteenth century, to add small quantities of indigo to their traditional vats to strengthen the colour.[69] Gradually more and more indigo was added to woad vats used for wool.[70] Meanwhile, vats of stronger alkalinity using only imported indigo were developed for linen and cotton. These used concentrated infusions of ash or potash, plus additions of large quantities of bran, madder, stale bread, molasses or glasswort.[71] One of the hardest things about dyeing with imported indigo was grinding up the lumps of hard pigment. Numerous dissolving substances and grinding mills were devised to facilitate the process.[72]

The Industrial Revolution in Europe caused intense research into better methods to mass-produce colourfast dyed and printed textiles; a bibliography by Lawrie lists innumerable dye manuals produced in the later eighteenth and nineteenth centuries which describe trials and recipes used for dyeing and calico printing.[73] Printing calicos became such big business that within the developing discipline of chemistry the profession of colourist became highly specialized, especially in Germany, France and Britain. Many experts debated the theories and practices of colourfast dyeing, for example the Anglo-American Bancroft,[74] the Frenchman Persoz[75] and the Scotsman Crum.[76] In his manual Bancroft devotes a huge section to indigo whose 'admirable and singular properties' are, he says, only just surpassed by those of shellfish purple.[77]

By the middle of the eighteenth century dyers of cotton, linen and silk were turning their backs on the slower warm fermentation methods thanks to the introduction of the first cold inorganic vat, namely the ferrous sulphate, or 'copperas' vat, which combined ferrous sulphate with slaked lime or potash. It was easy to set and produced good blues, but was expensive as it consumed much indigo and caused a rapid build-up of sediment. A century later the cheaper and easier zinc-lime vat was developed, which was based on slaked lime and zinc dust (which interacted to form hydrogen as a reducing agent), but it, too, produced unwanted precipitates and cloth had to be 'soured' in acid to remove the lime.[78] Both these vats were used throughout the century, but as they were strongly alkaline wool dyers stuck firmly to the so-called 'woad vats'. After much time and money had been spent on research, the reducing agent sodium hydrosulphite was introduced at the end of the nineteenth century.[79] First sodium bisulphite was combined with zinc dust and lime (which formed sodium hydrosulphite) in a vat which still produced precipitates,[80] but by 1904 a stable form of sodium hydrosulphite (dithionite), neatly coinciding with the new synthetic indigo, was introduced that heralded the modern indigo vat. This reducing agent combines with caustic soda or ammonia and can be adapted for use on most fibres. It also combines with both natural and synthetic indigo, consumes very little dye, does not produce sediment and is much easier to set and revive than the various methods it largely replaced.[81] It is still in use today, although research is intensifying to cut down on the use of polluting hydrosulphite and caustic soda (p.232). Some dyers still prefer the zinc-lime vat, despite the problems mentioned above, as it is cheaper and better for the skin, gives more solid, redder shades and is better for dyeing wax-resist cloth as there is less danger of loss of wax during immersion.

Resist-printed cloth, stretched on a star frame, being lowered into the indigo vat. Workshop of Josef Koó, Steinberg, Austria, 1995.

Meanwhile, woad was still used as a fermenting agent in indigo vats for wool dyeing (misleadingly called 'woad vats') even in the first decades of the twentieth century.[82] William Morris used such vats in the 1870s.[83] The practice seems to have been based more on historical attachment than on scientific necessity, for turnips have been shown to be just as effective as a fermenting agent. Indeed, in the nineteenth century rhubarb leaves sometimes replaced woad,[84] or chicory and carrot tops were mixed with woad.[85] Potato peelings were also explored as a possible fermenting agent which would be far more economical than woad.[86]

Horsfall and Lowrie give the recipe for an industrial woad-indigo fermentation vat, which survived into the twentieth century for dyeing loose wool for army khaki or airforce grey mixes, and for piece-dyeing serge for naval or police uniforms.[87] They noted that because of its mild alkalinity the vat was hard to control and demanded twenty-four hour supervision. The ingredients were:

1800 gallons water at 57°C
5 cwt woad
20–30 lb natural indigo, ground into a paste
5 buckets bran
6 lb madder
20 pints slaked lime

Over the next two days lime continued to be added as fermentation took place. On the third day the vat was ready for use, but only in the mornings, as daily additions of indigo, lime and bran took the rest of the day and night to dissolve and reduce.[88] Similar ingredients, without the woad, were used for a 'potash vat', while a sticky German 'soda vat' used soda crystals with lime, bran and much treacle.[89]

Today there are still endless permutations in the manufacture of an indigo vat, but dyers learn by experience what suits them best.[90] Even with synthetic indigo and modern chemicals it helps greatly to understand roughly what is going on, in order to keep the vat healthy and get even results. Otherwise, unforeseen accidents occur such as immersions in the vat that actually *remove* dye instead of adding it. For the craft dyer an indigo session is an incomparable experience, always full of fun. The late Susan Bosence, dyer and fabric printer (who infected the writer of this book with her enthusiasm), turned indigo-dyeing days into events to be shared with friends.[91]

Dyers' signboard of *Blaudruck* dye workshop in Hungary in 1866. On the left, loosely pleated cloth, on a circular metal frame that preceded the invention of the star frame, has been raised out of an indigo vat.

The dyeing of yarn for blue denim warp threads, or piece dyeing in bulk, is altogether a more serious business. To ensure even results (even harder in the case of piece dyeing), continuous dyeing ranges have been developed from a nineteenth-century system.[92] Warps or cloth pass over and under a series of guide rollers in and out of the tanks of dye, allowing time for full oxidation between each immersion.[93] Other more space-saving systems which mechanically feed yarn or cloth in and out of a single dye bath are also in use. In the past many methods were devised to ensure even dyeing and efficient oxidation. The last 'blue-print' dyers of Europe, for example, are still using ingenious metal star frames which hold many metres of printed fabric at one time – all 'cobwebly' as one guide book puts it – and prevent the faces of the fabric from touching each other while the frame is lowered and raised in and out of the vat on a pulley.[94] In Japan resist-patterned cloth is ingeniously tensioned on bamboo struts to prevent damage to the paste-resist.

Moon and Turtle

A Japanese dyer has claimed that natural and synthetic indigo are as different as 'moon and turtle', or, put the English way, 'chalk and cheese'.

What is the truth regarding synthetic versus natural indigo? Are they the same or different? The answer, to be paradoxical, is 'both'. The indigo blue is identical in both, therefore there should in theory be no visible difference. However, some people claim that cloth dyed in natural indigo has a density, like the 'body' of a fruity wine. This may be due to the repeated build-up required to obtain deep shades in a fermentation vat. Protagonists of natural indigo also claim that synthetic indigo is too uniform, whereas the so-called 'impurities' or 'contaminants' which occur unpredictably with natural dyeing provide variations beloved by many, including devotees of oriental rugs, who appreciate an uneven look, known as *abrash*. Such 'impurities' are caused by various factors. For instance, a reddish tinge can appear when indirubin is present in the vat (especially when fresh leaves are used).[95] One chemist who championed indirubin a century ago described its role as that of a 'passive assistant'.[96] Maybe the common addition of madder to indigo fermentation vats imparted a desirable red element. Sometimes yellow elements are caused by flavanoids.[97] This is most noticeable with woad-dyed wool, which tends to acquire a greenish tinge when left in the vat for long periods rather than repeatedly dipped and aired. In addition to 'indigo brown' and 'indigo gluten' there could well be other 'contaminants' such as tannins; and bacteriological elements as yet undetected may be at work in an organic vat, giving dyed fibres a livelier colour.[98] When used for medicine natural indigo appears to be essential, for recent tests indicate that it may be the presence of substances like indirubin that are most effective, rather than the indigo itself. There is clearly more to be learnt about this elusive topic. However, for most people the synthetic substitute produces results virtually indistinguishable from those of the natural dye, so that all but the real purist have turned their back on time-consuming organic vats made from natural indigo. Only deep-rooted cultural, economic or ecological considerations can save the organic indigo vat, symbol of an almost mystical closeness with the natural world.

'Finishing' indigo-dyed cloth

Beating indigo-dyed cloth with heavy wooden mallets to make it stiff and shiny. Zabid, Yemen, 1989.

Newly dyed indigo cloth, although colourfast, has a tendency to surface rub-off. Though disliked in some societies, this is much appreciated in others, either because it is considered to have medicinal properties (see p.222), or because the imparting of a blue tinge to the skin from the coppery surface sheen of beaten cloth bestowed prestige, if not an almost spiritual aura, on the wearer. An English traveller, Clapperton, at Kano in Nigeria in the 1820s noted that the cost of having a robe dyed was 3000 cowries, but it cost 700 more cowries to have it glazed.[99] A shiny robe, turban or face mask certainly looks most impressive when worn in the sun. In other societies dyed cloth has been treated in most ingenious ways to render it almost waterproof and reduce undesirable rub-off when, for example, used as a background to embroidery. In the West in the nineteenth century ingredients such as fuller's earth and urine (again) were sometimes used commercially to help prevent 'rub-off'.[100]

Treatments to 'finish' cloth have included physical polishing by calendering or burnishing, or the application of glazing substances, or a combination of the two. In India cloth was 'beetled' with wooden beaters and 'chanked' with smooth shells, and this practice may have spread to southern Arabia and thence been transmitted with Islam to West Africa. In southern Arabia indigo-dyed cloth is still occasionally coated with starch, gum arabic and neat indigo solution before being beaten with heavy wooden mallets and in some cases also burnished with a special stone (a practice also recorded in Nigeria in the 1960s)[101] in a locked room to prevent the djinns from spoiling the work.[102]

In West Africa newly dyed cloth is often beaten with wooden beaters to iron it, and when the sheen wears off cloth will be returned to the beaters. For Muslim groups living around the central and western Sahara, notably the Touareg, much more effort went into creating a sheen as iridescent as carbon paper on special narrow-strip indigo cloth known as *turkudi*. The manufacture of this cloth has dwindled greatly, with only a few dyers still producing it near Kano, but in the past it occupied the indigo dyers of

Elderly man proudly wearing a shiny indigo-dyed turban with fresh herbs stuck in the side. Tawilah, Yemen, 1981.

Cameroon and Chad as well as Nigeria.[103] Clapperton, the traveller mentioned above, noted that whenever a new log for beating on was brought to the gates of Kano city it was ceremoniously accompanied by drummers and attracted a large crowd.[104] His companion, Denham, travelled to western Chad, where he admired the high sheen on local indigo-dyed costumes, produced (as in Nigeria) in the following manner with antimony, or kohl:

The [indigo-dyed] linen ... is laid in a damp state on the trunks of large trees, cut to a flat surface for the purpose, and are then beaten with a wooden mallet, being at the same time occasionally sprinkled with cold water and powdered antimony; by this means, the most glossy appearance is produced: the constant hammering attending this process during the whole day really sounds like the busy hum of industry and occupation.[105]

Travellers in Yemen reported hearing the same continuous din 150 years later.[106]

This popular 'bronzy' effect still so beloved in West Africa is today created commercially in a French factory in Mali using friction produced by two rollers moving at different speeds, on indigo cloth dyed with less hydrosulphite than normal to minimize reduction of the dye.[107] As methyl violet was formerly used on cloth dyed in England to emulate the bronzy effect for the West African market,[108] it is fascinating to find dyers both in West Africa and south-west China doing much the same thing today. In both areas indigo or other blue-dyed cloth is given a final coating with a bronzy-purplish dye akin to Gentian Purple (known in West Africa as 'Mali Purple'), which, when beaten on, simulates the indigo sheen.[109] In parts of West Africa a woman will still check for real indigo by smelling and rubbing cloth with her hands, knowing at once that the indigo is not genuine if the rub-off is purple. Today, however, the general passion for shininess is more easily satisfied by modern lurex.[110]

In south-west China dyed cloth is still hammered hard with large wooden mallets. It is also calendered, as silk was in the past elsewhere

Touareg man of the Aïr mountains beside his tent, wearing a prestigious *tagelmoust* of indigo-dyed narrow strip cloth with extra indigo dye pounded in. Saharan village of Kouboubou, Niger, 1970.

Calendering indigo-dyed cloth using a large wooden mallet. Biashia, a Miao ethnic minority village, Guizhou, south-west China, 1993.

Calendering dampened indigo-dyed cloth, wound around a wooden roller, by standing on a huge arc-shaped stone. Buyi minority village, near Zhengfeng, Guizhou, south-west China, 1993.

in the country, in an extraordinary way. It is wound on a roller and polished by the rocking motion of an enormous arc-shaped stone, the dyer either standing on the stone and tilting his body from side to side, or rocking it by hand. This resembles a contraption still used in isolated parts of central Europe to confer a chintz-like effect on 'blue-printed' cottons in emulation of the popular Indian glazed cottons. Here the polishing of the indigo-dyed cloth on its wooden roller is achieved by means of an enormous box mangle filled with ballast (a giant version of the former box mangle system of ironing widely used in Europe). Originally the mangle was hauled back and forth across the cloth using horse power, but it is now diesel-driven.

The minority tribal people of China, northern Thailand and nearby regions apply and beat in, or steam-fix, many substances as glazes, some of which would further darken the cloth, too. Several of the substances sound more appropriate to the cook than to the dyer, including as they

Lengths of newly dyed indigo cloth hanging ove the village balconies to dry prior to being 'gummed'. Goahzen village, Dong minority area, Guizhou, south-west China, 1993.

Calendering 'blue-print' cloth, wound around wooden rollers (taking place on the right of the picture) by the friction caused when a huge box-mangle was hauled to and fro over the rollers using horse-power. Dyers' sign-board, Hungary, 1862.

Applying gum made from soaked tree bark to indigo-dyed cloth to give it a surface shine. Goazhen village, Dong minority area, Guizhou, south-west China, 1993.

do bean starch, red peppers, egg white and persimmon juice (with its waterproofing qualities and its natural tannin).[111] Different localities use whatever gums or animal proteins are available: home-made ox-hide glue, the juices of soaked tree barks (the inner bark layer contains starch, gums and tannins),[112] roots or leaves, and even ox or pig blood. Karo Batak indigo dyers of northern Sumatra similarly combine bark gums with buffalo blood.[113] Blood glues are so water-resistant that as late as 1979 they were still being patented in the USA.[114] The iron content of blood may also react with cloth to make it darker. The Chinese have even smoked their cloth as if it was fish.[115]

Dyers' guilds, status and sex

Because of the difficulty of mastering this peculiar craft, indigo dyers worldwide have been specialists, jealously guarding their secrets and passing them on from generation to generation, deliberately creating an aura of mystique around the whole process.

It is impossible to make dogmatic pronouncements about the status of dyers as a whole as there are so many variables to consider. At one end of the spectrum we find that dyeing with precious shellfish purple was in ancient Palestine the prerogative of members of the priestly class, while at the other end any woman could handle simple yellow dyes at home. Between these two extremes fall the indigo dyers, whose status varied from country to country. However, even in societies where dyeing in general has been perceived as a rather messy and lowly occupation, indigo dyers have received a measure of respect, as their particular skills have been recognized as setting them apart from other dyers. The give-away sign of fingernails stained permanently blue – Mexican dyers were sometimes known simply as *tecos* ('finger nails')[116] – was nothing to be ashamed of.

In the Western world skilled craft workers since Classical times have formed themselves into guilds or associations to protect their livelihoods.[117] Dyers have been no exception, and indigo or woad dyers usually formed independent associations within the overall guilds of dyers or cloth-workers. Dyers' associations, which seem to have been in existence in Ptolemaic Egypt, were to be found throughout the Roman Empire. The status of dyers in Pompeii is indicated by their substantial meeting hall and workshops with their associated statuary. Even in those days dyers with the skills required to handle colourfast dyes were highly regarded. Despite their specialization, some indigo dyers apparently used red dyes as well, probably because of the demand for imitation shellfish-purple colours in Roman and Byzantine times. This would explain the presence of red-stained vats amongst those unearthed in the Egyptian Roman indigo-dyeing workshop mentioned on p.24. By the Middle Ages the textile industry generally, both in Europe and the Islamic world,[118] was being subjected to increasing regulation, which provided protection for the guilds and also generated revenue for governments.

In the golden age of Islamic textiles indigo dyeing was often in Jewish hands.[119] Their status is hard to pinpoint. Although not as esteemed as the dyers of real shellfish purple, indigo dyers were indispensable from antiquity. Jewish dyers upheld their position in the Middle Ages and beyond, hence the common surname Sebag or Zebag, from *sabbagh*, meaning dyer. They were prepared to pay heavy taxes to retain a monopoly on their trade. When Arabs swept into Sicily and southern Spain Jewish dyers from the

Indigo-dyed cloth drying in the sun on a hillside after the application of bark gum by the dyer, who is wearing everyday village dress of indigo jacket, pleated skirt and puttees. Goazhen, Guizhou, south-west China, 1993.

Arab world brought the art of indigo dyeing into these areas, and were probably influential in teaching Spanish and Italian dyers, who were the first Europeans to switch to tropical indigo, how to use it.[120] By the eighteenth century Muslims in many parts of the Arab world were also dyeing with indigo, but the specialization remained strong, and the pride taken in such dyeing is expressed in the words of a Yemeni writing in the eighteenth century: 'Glory lies in learning and a hand black from the craft of dyeing, not in the company of Lords.'[121] In central Asia the so-called 'cold' (i.e. indigo) dyers of tied yarns for the famous ikat silks were always Jews, while 'hot' colours were dyed by Tadjiks.[122] In Persian Isfahan the indigo-blue cotton cloth, *qadak*, worn by all classes was manufactured by an important guild. Even though European fabrics of assorted colours appeared there in the 1860s, in the 1870s there were still 136 elegant dye shops along the lively *qadak* bazaar.[123]

In medieval Europe, too, wealthy dyers and entrepreneurs who specialized in woad, the major wool dye, often enjoyed a higher status than others in the textile industry, their profession being no bar to becoming city mayor for example.[124] (The exclusion of people with 'blue nails' from membership of the Merchant Guild of Edinburgh in the twelfth century is therefore surprising.)[125] A big distinction was made between dyers who handled the 'lesser', *petit teint* dyes, using more fugitive, simpler colours on cheaper cloth, and those who had mastered the expensive colourfast, *grand teint* dyes such as woad, madder and kermes.[126] Sometimes German woad dyers had to undergo stiff examinations to prove that they had fully mastered their skills.[127] In Italy letters and recipes in the archives of the Datini dyeing company of Prato, and recipes in the famous fifteenth-century Florentine *Trattato d'Arte della Lana*, express the particular pride taken by guild master dyers who could dye wool really well with woad.[128] Woad dyers were instructed to keep the tricks of the trade strictly to themselves,[129] and were heavily fined if they moved to another town and divulged their recipes.[130] Everywhere in Europe ordinances and prohibitions were constantly being issued in an attempt to control the industry and maintain standards among dyers.

Many changes were initiated during the Renaissance. The invention of printing soon brought to life books of printed dye recipes, appropriately termed 'secrets', the first significant one being Rosetti's *Plictho de larte de tentori* . . .

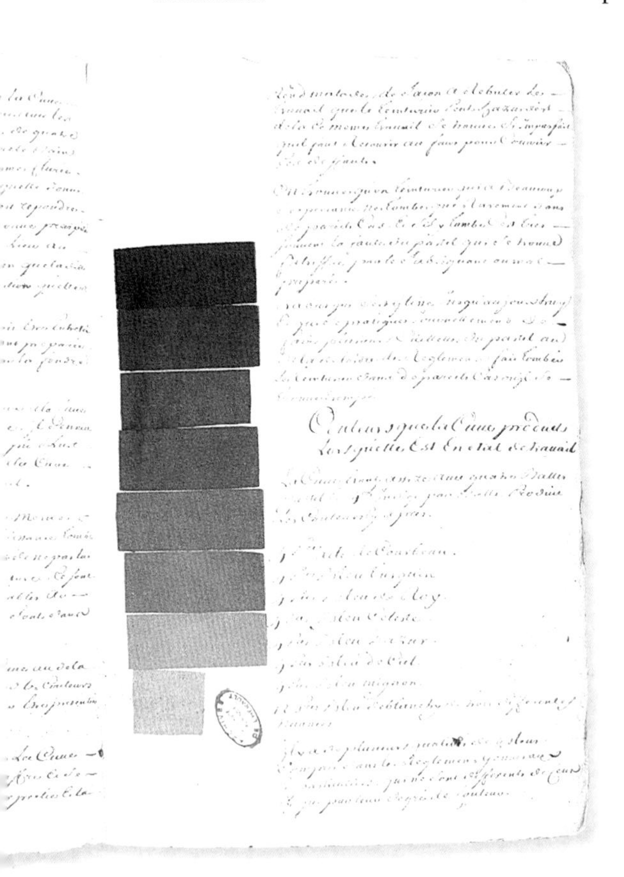

Sample book of 1744 showing eight named shades of blue produced on woollen cloth in the woad/indigo vat by dyers of Languedoc, France. Names range from 'crow's wing' to 'almost-white'.

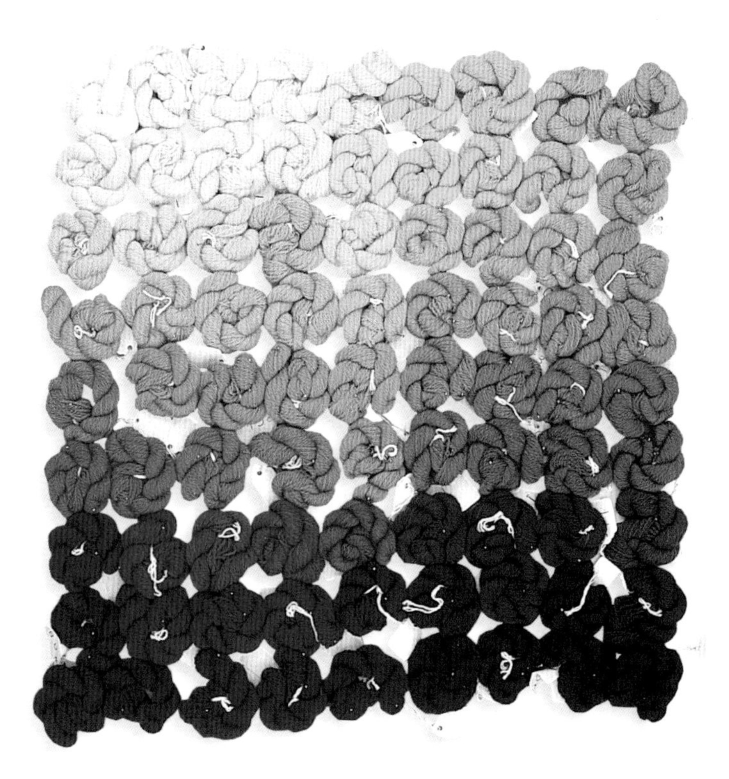

Woollen samples showing shades achievable using only indigo and yellow weld. Dyed by Margaret Redpath for 'Renaissance Dyeing', Wales, in 1998.

of 1548.[131] This manual, with its detailed descriptions of solid and fugitive colours, opened up the closed and secretive world of the dyers, and placed dyeing on the first rung of a more scientific ladder at a time when exotic new dyes like indigo were becoming more widely available.[132] In the same century and the next the British and Flemish also began to apply a more scientific approach to dyeing and the selection of dyes. Typical Dutch municipal regulations in the seventeenth century permitted dyers to use woad for blue but forbade the use of inferior logwood. More important for the future of the dyeing industry was the royal edict drawn up in 1671 by Louis XIV's minister, Colbert. His *Instruction gènèrale pour la teinture des laines* … which was disseminated throughout Europe, aimed to standardize practices, promote colourfast dyes and prohibit inferior dyes of *le petit et faux teint* such as logwood. It also sanctioned adding some indigo to woad vats.[133] In the following century, thanks to the appointment of academicians to direct the Royal Manufacture of Gobelins, further efforts were made to understand how dyestuffs worked and what made them colourfast,[134] and scientists such as Jean Hellot published studies of woad (which he considered the most difficult of all dyes) and 'exotic' indigo.[135] As Bemiss' early nineteenth century dye manual reaffirmed, 'blue among all colours is the most difficult to set up and manage';[136] but this did not prevent the Gobelin dyers distinguishing by this time no less than sixty different shades of blue.[137]

Turning to other parts of the world, again we find indigo dyers holding a singular position *vis-à-vis* other dyers. In India the complex issue of status amongst dyers was especially pronounced. It was interwoven with the intricacies of the country's caste system, as well as attitudes to different textiles. The type of fibres to be dyed could dictate the social position of the dyer; weaving and dyeing of silk and wool, ritually pure fibres, could only be undertaken by higher-caste Hindus, whereas working with cotton, considered by Hindus to be a transmitter of pollution, was generally the domain of lower-caste Hindus and untouchables.[138] In north-western India and the Deccan, Muslims have dominated the dyeing profession, although as indigo dyeing pre-dates Islam, dyers must at one time have been

Batik sarong (detail) from the Pekalangan area, Java, early twentieth century. This darker section of the sarong is known as the 'head'.

Dyeing with local wild *Indigofera arrecta*. Kani-Kombolé village, beneath the escarpment, Dogon area, Mali, 1997.

Hindu and may even have embraced Islam to escape their lowly status. In western India cotton dyeing is still undertaken by Muslim dyers whose generic surname is *Khatri*, while indigo dyers elsewhere in India were often called *Nilari* or *Nilgar*.[139] Their specialization is especially apparent in the dyeing of India's famous chintzes. Cotton painters and printers were highly skilled in using a wide range of mordant dyes, but indigo dyeing was always sub-contracted out to independent specialists at the appropriate stage. To keep the secrets of indigo dyeing watertight within the family, Indian dyers were wary of passing them on to marriageable daughters.

In West Africa indigo dyeing has usually been the domain of certain ethnic groups. The subject of diffusion of skills is complex, but it appears that Soninké dyers from Upper Senegal migrated with their art of indigo resist-dyeing south and east into the Gambia, Mali, Sierra Leone and Guinea, as well as westwards to St Louis, where they influenced the Wolof peoples.[140] Certainly, even today many indigo dyers of the region are of Soninké (Sarakholé, Maraka) origin and Soninké terms for indigo dyeing are still found throughout the region. The Baule around Bouaké in Ivory Coast were renowned indigo dyers, and in the Cercle de Bondoukou, straddling Ivory Coast's frontier with Ghana, indigo dyeing is in Muslim hands as a result of the spread of Islam.[141] The indigo dyers, also Muslim, of Garoua in Cameroon were of such repute that people were prepared to undertake journeys of over 1500 km to have their *ndop* cloth dyed by them.[142] In Nigeria indigo dyeing has been in the hands of Yoruba and Hausa people. In Yorubaland the dyeing of the famous *adire* cloth was restricted to women, men and boys being kept from the dye pots at arm's length, whereas in the north of the country the Muslim Hausa dyers, notably of Kano, were always male.

Dogon indigo dyers of Mali typify the ambivalent attitudes of communities towards the profession. They belong to the caste of artisans who work with substances that have a high, though potentially destructive, spiritual energy. The caste includes sculptors, special 'praise singers' and leather-workers, who, like indigo dyers, undertake work that is both 'impure' and yet technically skilful.[143] In Sierra Leone the status of indigo dyers has always been high. If a woman was not passing on the secrets of indigo dyeing to her own daughter, she could demand a high fee in goods for teaching a kinswoman or a friend. In the past in Sierra Leone it was said that 'the mark of a free woman was that her hands were always black with dye'.[144]

The sex of indigo dyers has varied according to local custom and demand, but the general tendency worldwide has been for men to carry out the dyeing on the larger commercial scale, not surprisingly given the traditional division of labour. Sometimes the mere presence of a woman

Indigo-dyeing at Zhaoxing, a Dong minority village, Guizhou, south-west China, 1993.

near an all-male dye-house was considered inauspicious in the Middle East.[145] On the other hand, it is still women who tend to dye at home on a domestic scale, especially in places with deep-rooted beliefs and superstitions relating to indigo. Nonetheless, they are still specialists within their village, or sometimes within a wider community, whose weavers would bring them yarn for dyeing. One example is that of Tenganan in eastern Bali, where women have been famous for their skills in weaving sacred double-ikat cloth, *geringsing*. Here, although they dye the morinda red themselves, they do not tackle the indigo, but instead take the prepared yarns to a neighbouring village for this stage in the production.[146] On other Indonesian islands, and in other parts of Southeast Asia, most women would learn to tie yarn for ikat and to weave, but mastery of the difficult dyes, morinda and indigo, was the most highly respected textile art of all.[147] Only a chosen few would be initiated into the secrets of such dyeing, passed on down the maternal line.[148] Among Khmer people in southern Thailand an indigo dyer will only pass on her secrets after her intended successor has performed an elaborate 'respect ceremony' involving an offering of a flower bowl filled with a strange combination of items, and a thorough dousing with powders.[149] Such protective secrecy can be a disadvantage, for traditional indigo dyeing sometimes vanishes, secrets and all. Many indigo dyers have been convinced that betraying their profession to an outsider would bring dire misfortune to their family; one researcher on Javanese textiles comments wryly: 'The mixing of the yearly supply of *nila* is shrouded in the highest secrecy . . . in my seven years in the area I have not been allowed actually to see the complete sequence of preparation.'[150]

Dyeing ten-metre lengths of calico in the last indigo workshop, Bayt Muhammad Ali Abud, of Zabid. On the Tihamah, the Red Sea coastal plain of Yemen, 1983.

Deep within the Japanese psyche is a special empathy with indigo. Before the twelfth century most women would dye their families' clothing themselves, but thereafter the independent male profession of indigo dyeing developed in both cities and villages.[151] The status of indigo dyers has for centuries been high in Japan and today they are even nationally recognized as 'living treasures'. Perhaps we should cherish in the same way all the world's remaining dyers of natural indigo before it is too late?

You have to work in partnership with indigo,
you cannot impose upon it
(BOBBIE COX, 1998)

The Variety of Decorative Techniques

The relationship between dyes and cloth, so different from that of paint and its underlying surfaces, is infinitely subtle. In the words of a celebrated Japanese dyer, Shigeki Fukumoto: 'Our [the Japanese] dislike of paint is the reverse side of the coin to our love of dye.'[1] Each dye has its own characteristics but indigo's unique bond with the widest possible variety of natural fibres creates qualities hard to define but easy to appreciate, hence the passion many people feel for undecorated indigo-dyed cloth. However, people all over the world relish patterned clothing and furnishings, which can be the most colourful element in their lives. Since finely woven and embroidered textiles were luxuries often out of reach of the ordinary person, over the centuries various methods were devised to create patterns more easily, thereby making decorative textiles available to all. Inevitably some of the techniques developed into elaborate procedures in their own right, but others have remained bold and simple. Indigo has a long-standing link with many of these techniques since in most places, particularly rural areas, it was used for dyeing yarn or cloth patterned with resists. The result was typically an indigo-blue ground with white patterning, always an appealing combination.

People who pattern cloth with indigo praise its ability to spring surprises. Indigo compensates for certain limitations to do with the fact that it is not a mordant dye by offering a huge range of decorative possibilities of its

Narrow strip cloth (detail) crumpled and tied in a bundle before dyeing to produce a marbled 'cloud' pattern. Collected in the 1860s in the Gambia or Liberia.

Ajrakh turban and shoulder cloth dyed in indigo and alizarin red, worn in western Kutch, where, as in neighbouring Pakistani Sind, it is a mark of Muslim culture in the region. India, 1994.

own. With hot mordant dyes, such as madder, patterns are usually created by a combination of mechanical and chemical means. A thickened mordant, commonly alum, is printed or painted onto cloth where a coloured pattern is desired. Subsequently, in the hot dye bath the mordant bonds chemically with the dyestuff and the fibres to form the coloured elements in the design. Alternatively, some of the cloth can have a mechanical resist like wax or mud applied which will resist a cold mordant paste. The wax or mud resist can then be removed before applying a hot mordant dye. Either way the dye will only fix to the parts treated with mordant. The complexities of resist-patterning with mordant dyes are unravelled by Ruth Barnes in her study *Indian Block-Printed Textiles in Egypt*.[2]

As indigo works in the reverse way to mordant dyes, the concept of resist- or reserve-patterning is easier to grasp. Dyers using indigo make a virtue out of necessity by generally creating designs by mechanical processes and subsequently putting the cloth in the dye vat, rather than bringing the dye to the cloth. Parts not intended to be dyed must be reserved in some way before immersion in the dye vat, so that they will resist the dye. As indigo is a tepid dye, and immersions in the vat are short, patterning techniques can be employed which involve the application on areas to be reserved of delicate substances such as wax and dried pastes which would melt or disintegrate in hot dye baths or long immersions. Furthermore, with other resist techniques of tie-dyeing or stitching (*shibori*), the nature of indigo's oxidation process and the build-up of colour enable literally endless nuances of shade to be produced.

Indigo's limitations become apparent when it is required for printing or painting the positive, i.e. for blue motifs on a white or other background (although the illusion of a positive design in blue can be created by skilful mastery of blocks and hand-painted resists). There are, however, two partly chemical methods of patterning with indigo, developed in Europe in the eighteenth and nineteenth centuries. One is the 'discharge' method in which the pattern is bleached out of cloth already dyed in indigo; the other (known as 'Pencil Blue' and 'China Blue') involves painting or printing directly with indigo. The term 'blue-printing', however, is misleadingly used. Although one would expect it to apply to printing with blue, it is in fact used for the quite different central European *Blaudruck* ('blue-print'), which is a white-on-blue block-printed resist process.

The various ways to pattern cloth with indigo will here be broadly divided into two categories, distinguishable by the visual results: those where the pattern is quite sharply defined and those where its edges are characteristically blurred by manipulating the cloth or yarn. A useful and lavishly illustrated survey of all the resist techniques is *The Dyer's Art*,[3] and Alfred Bühler's more comprehensive three-volume work on the subject remains a classic study.[4]

Miao minority child's skirt with wax-resist and embroidered panels. From Biashia, Guizhou, south-west China. Collected 1993.

Hard-edge patterning

Sharper definitions appear when parts of the cloth not intended to be dyed are covered before dyeing by the application of various substances such as hot wax or cold pliable paste, or are reserved with devices like clamps. The texture of the base cloth greatly affects the end result, but generally the more a resist penetrates the cloth fibres, the sharper the pattern will be, as the dye will be more completely repelled.[5] The resists themselves have their own special characteristics, such as the distinctive fine 'crackle' often found in hot wax-resist batik, or the softer marks made by clay and paste resists. The creative possibilities of resist-dyeing are endless. Modern Japanese artists in particular excel at exploiting these possibilities to produce the subtlest results.[6]

The term 'batik' is associated with Javanese hot wax-resist methods although it is sometimes applied collectively to all kinds of resists which are applied to cloth before dyeing and then boiled or scraped off afterwards. All the various types of resist can be applied freehand, or using blocks or stamps of some kind, or through stencils. The last two methods usually came later to speed up the laborious nature of hand-drawn work. The most well-known resist medium is hot wax, usually pliable beeswax or a similar resin, or modern paraffin wax. Hot wax penetrates cloth exceptionally well, and its heat and suppleness can be tailored to suit different types of fabric. Apart from wax, other substances applied to fabric as resists are the starchy pastes made from flour and beans, pipe clay, mud and animal fats. The more fragile resists are best suited to the milder indigo fermentation vat.

No one knows the origins of batik, but textiles found at sites in China, Egypt and central Asia (some are mentioned in Chapter Two) show that it has been practised in different parts of the world for over 2000 years.[7] Nearly all the earliest examples extant were dyed with indigo. Other places known for batik include Java, India, Sri Lanka, Japan, Vietnam, Thailand, Turkestan and West Africa. Until synthetic dyes became widely available, especially in the second half of the twentieth century, indigo remained the commonest dye used for most resist-printed and painted fabrics, wherever they were produced.[8] Today the art of batik is found worldwide, thanks mainly to an enthusiasm for it sparked off by exhibitions of Indonesian batiks first held in Europe in the late nineteenth and early twentieth centuries.[9] However popular its practice may have become in the West, it is to be hoped that it will remain a useful, low-cost, income-generating medium amongst craftsmen in the developing world.

The East

The Javanese elevated batik to a high art and are probably responsible for the word 'batik' itself. Both in Java and in some other Indonesian islands, notably south-east Sumatra, batik production was an important domestic industry, the patterned cloths playing a major role in local costume as well as being widely traded, particularly by the colonizing Dutch.[10] In Java the wax is applied in two different ways. In the first, known as *tulis* (meaning 'writing'), patterns are painstakingly hand drawn by women using a *canting* (or *djanting*). This bamboo- or wood-handled instrument has a small reservoir, usually of copper, out of which hot wax flows through a spout or group of spouts. Very precise line work can be produced, especially when applied to cotton or silk sized with rice starch. In the mid-nineteenth century, in response to competition from imported European machine-printed batik imitations, a speedier method was adopted which involved men printing hot wax with pairs of large stamps, called *cap* (*tjap*), usually made from sheet copper.

Hand-drawn batik sample from Java with elephant design in two blues (23 cm square). Early twentieth century.

Raffles, during his governorship of Java in the early nineteenth century, was the first European to record and publish the batik process in detail.[11] A famous colour combination produced in central Java is that of the 'soga batiks', whose white or cream ground contrasts with indigo blue and a rich tan colour (which gradually fades) made from the bark of the soga tree. Great care was taken to prevent any 'crackle' effect in the indigo by warming the wax in the sun before dyeing; but in parts overdyed in soga the wax was deliberately broken up to create an overlay of brown marble-like veining.[12] Blue-and-white monochrome batiks were particularly favoured by Chinese and Dutch buyers.

In China, blue-and-white batik was popular during the Tang dynasty (AD 618–906) but at some stage hand-drawn batik seems to have died out

)
nese hand-drawn 'soga batik'
digo and soga brown.

ow left)
go-dyed wax-resist jacket and
with a set of tools used to apply
wax, and a lump of recycled wax.
n by Buyi minority women, Shittou
ge, Guizhou, south-west China.
ected 1993.

among the Han Chinese.[13] However, people of several ethnic minorities living in China's south-west provinces and the island of Hainan, and across her south-western borders, are still well known for their indigo-dyed hand-drawn batik.[14] Various groups among the Miao/Hmong are particularly famous for such work, called *laran*. It is used for skirt panels and the all-important baby carriers, as well as for panels inserted in other garments and furnishings.[15] Batik costumes are produced by girls in preparation for marriage, and their textile skills are shown off to best advantage at festivals, where they attract suitors. The resist mediums used (on cotton today but on hemp in the past) are usually hot wax or resin, including maple sap. One researcher has recorded the use of pigs' fat to make the mixture more supple,[16] and such tallow was also used in central European block-resist paste recipes.[17] Simple home-made tools include brushes made from hairs fixed to a quill handle with a blob of wax. Yet with only such basic tools at their disposal, women master resist application (often on both sides of a starched cloth) in such a way as to leave a fine line of indigo between areas of resist that give an impression almost like blue ink drawing. For the tourist market, however, standards are lowered: the narrow cloth that traditionally forms the border of a pleated skirt, for example, is fashioned for foreigners into a purse or bag instead.

Miao (Hmong) woman applying hot wax before dyeing the strips in indigo. Pua, northern Thailand, 1992.

Instead of hand-drawn batik the Han Chinese developed a method of applying cold paste-resists using a stencil. These rural cloths, too, were dyed in indigo, producing lively blue-and-white designs especially popular in central China.[18] In both China and Japan stencils are cut with great skill out of mulberry paper waterproofed with tung oil or persimmon juice until it resembles parchment. Instead of wax, a resist paste is applied with a knife through the cut-out parts of the stencils. In China this is made from lime and soya bean curd and in Japan from rice flour and lime, but in Djakarta, where such indigo-dyed stencil work was formerly common, crushed groundnuts were used.[19] The paste must dry hard before the cloth is dyed.

Hiroshige colour woodcut (no. 19 of 36) of a seashore scene, 1858. The fisherman is wearing an indigo-dyed kimono with stencilled designs.

Nigerian *adire eliko* cloth. In the piece underneath, patterns have been applied with cassava flour; above is a completed cloth.

In Japan, as elsewhere, imported Indian printed and painted cottons, known there as *sarasa*, became enormously popular in the seventeenth century. Japanese dyers, who were already familiar with stencil resist-dyeing, *norizome*, soon began to produce their own versions of Indian cottons. Today they continue the art of applying paste with stencils, *katazome*, or squeezing it through a cone, *tsutsagaki*. The Japanese elevated paste-resist work to a highly refined art, using indigo blue as the classic ground colour as well as exploiting its nuances of shade.[20] Garments made from all types of stencilled cloth feature in Japanese woodblock-printed pictures. The Japanese also had an ancient tradition of wax-resist dyeing, *rozome*, in the seventh century AD, but it died out after a hundred years and was not resurrected until the twentieth century.

Since the Second World War batik factories have sprung up in many places in the Far East. They often produce cheap, crudely executed batiks using modern chemical dyes of all colours. Industrially printed imitations are equally common. However, the classic blue-and-white colour combination retains a timeless appeal.

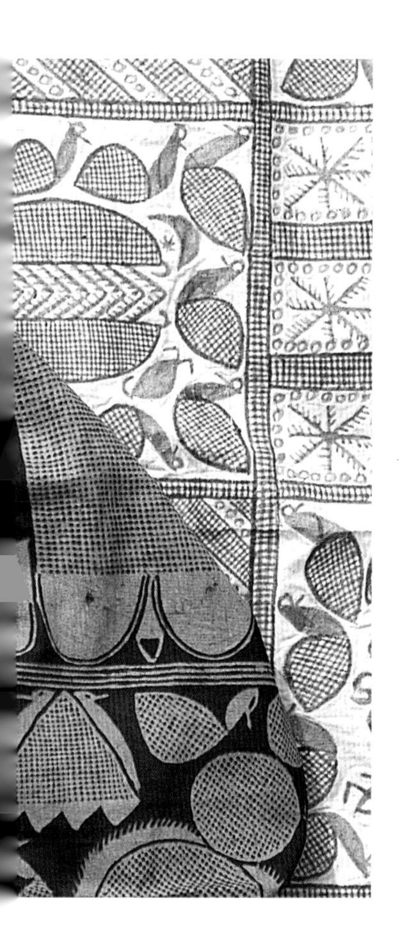

African paste- and wax-resist

Also famous for paste-resist work are the Yoruba of Nigeria. The general Yoruba term for resist-patterned indigo-dyed cloth is *adire*, while *adire eliko* refers specifically to cloth patterned by applying a starch resist, either free-hand with a comb or through a metal stencil. All types of indigo-dyed *adire* are well known in the West, their lively designs having made them popular, and there are large collections in many museums. An article written in the 1930s records that almost half the population of Yorubaland was then involved in the indigo industry, with over £200,000 worth of cloth being dyed annually. There were several production centres, but cloth from Aboekuta, 'the Paris of Yorubaland', was the most renowned and had a thriving export trade to neighbouring countries.[21] Today the art is being kept alive in Yorubaland by the Nike Centre for the Arts at Oshogbo.

The paste used for *adire eliko* is made from the local glutinous cassava flour, conveniently to hand as it is a common cooking ingredient. Women paint on the paste, to which alum and copper sulphate (known as 'blue alum') is added, as it is to central European paste-resists. Blue alum colours the paste, making it easier to see, and is said to help repel the indigo by chemical means. Traditional painting tools are small chicken feathers, sticks or palm leaf ribs.[22] As elsewhere, in Yorubaland stencils were developed in the early twentieth century to speed things up, and using them is men's work. Unlike the very refined Japanese mulberry paper stencils, Yoruba stencils were formerly made from lead or leather, and now zinc or iron sheeting, in bold designs.

Yoruba people prefer the appearance of pale blue against dark blue, rather than the starker contrast of blue and white. This is probably because paste does not penetrate cloth as fully as wax, so that some indigo colours the reverse of the pasted areas of cloth, giving an appearance of pale blue when paste is removed. The blue-on-blue effect was usually enhanced by giving *adire* cloth a final dipping in indigo after removing the paste.

Hot wax seems to have appeared in West Africa within the last fifty years and is widely used, often applied with wooden stamps.[23] In Senegal an unusual method of combing designs through a rice paste was recorded in 1910. These combed cloths were dyed in indigo, and a similar technique using wheat starch can be found in Sierra Leone and the Gambia today, but no longer with indigo.[24] The famous mud cloths of Mali, laboriously made by applying mud (to pre-mordanted cloth) *around* the patterns to create a resist-patterned appearance of pale designs on a dark ground, may well owe a debt to the negative-on-positive design of indigo-patterned cloth so characteristic of the whole region.

'African print' made recently for the West African market by A. Brunnschweiler and Co. (ABC), near Manchester.

Factory-made 'African' and 'Fancy' prints

The story behind the production of 'African prints' in the Netherlands and England provides a fascinating vignette on the ripple effects of colonialism.[25] In the mid-nineteenth century the Netherlands started to manufacture imitation Indonesian wax prints intended for export to Java. However, demand there soon declined when local production methods speeded up thanks to the invention of the *cap* (see p.151). The Dutch therefore sought other markets for their factory-made batiks, and hit on the Gold Coast (Ghana), where a taste had already developed for Java batiks, partly thanks to returning mercenaries recruited there by the Dutch to help them control Indonesia. The appetite in Ghana for Java-style batiks was also fed by Dutch traders and European missionaries, and the fashion for these 'African prints' spread throughout West Africa. They were manufactured in factories in the Netherlands, Switzerland and England. To this day Vlisco Company in the Netherlands and A. Brunnschweiler and Co. (ABC) in England,[26] still manufacture genuine wax prints, aimed at West African markets, by applying mechanically to both sides of the fabric a resin that creates a wax-like crackle. Until fifteen years ago almost all of ABC's 'African prints' were dyed in indigo. In Ghana itself, where two genuine 'wax-print' factories operate, most of the millions of metres produced each year for the home market are still of classic indigo, the Ghanaian colour of love, and white. Cheaper imitations are also widely produced today by factories in Africa and Asia. These 'Fancy' prints are printed on one face of cloth only in all colours.

Indian printed and painted cottons

The parts of India for centuries world famous for painted and printed cottons are the south-east and the north-west. Their trade descriptions refer to them as patterned cloth, but were applied indiscriminately to both painted and printed varieties. The Portuguese used the term *pintados* (painted). Other European terms, derived from the Hindi *chitta* (spotted cloth) or *chint* (to paint), were *chintz* (English), *chittes* (French), *sits* (Dutch) and *zitz* (German). These cotton cloths were multi-coloured. The combination in them of mordant dyes with vat-dyed indigo challenged skilled Indian dyers to the full.

In order to produce red and blue in handblock-printed or painted textiles, and to avoid an overlapping purple where not desired, dyers have to be able to juggle mordants, resists and dyes. Reds and blues appear together in the block-printed cloths of north-west India, which have been traded far and wide, at least since the early Middle Ages.[27] *Ajrakh* cloth is still today a mark of Muslim Sindi culture both in Pakistan and in Indian Kutch and western Rajasthan. A traditional *ajrakh* is characterized by its geometric designs whose colour scheme is alizarin red and indigo blue; indeed, the word *ajrakh* is a corruption of *azraq*, the Arabic and Persian for blue, the predominant colour of these cloths. *Ajrakhs* function as all-purpose cloths. They are used by men as turbans and shoulder-cloths, and by women for shawls, hammocks, etc. Producing an *ajrakh* by hand is a long and arduous job whose process differs from centre to centre. Wooden printing blocks are hand-carved into intricate patterns by specialists, and many designs have been passed on down the centuries.[28] To prepare fibres for dyeing, cloth is elaborately steamed, softened and bleached with substances such as oil and camel dung, or modern soda. Then follows a complicated sequence of printing with mordants and resists. These can be used in a variety of ways, separately or combined, and the indigo blue can either be dyed before or after the alizarin red, depending on the methods chosen by the dyer. Many variations are recorded in Bilgrami's book *Sindh Jo Ajrak*.[29]

The resist medium commonly used both for *ajrakh* and for the printed cloths of eastern Rajasthan is a clay-rich mud. Recipes are a well-kept secret, the main objective being to make the mud adhesive enough, by adding substances such as gum arabic, castor oil, lime and wheat flour, to withstand the indigo dye bath, kneading it with the feet until the consistency is right.[30] It is much easier to create the colours red and black by printing with mordant dyes, notably alizarin and iron, and omit indigo blue, as is still done for cheaper cloths in Ahmedabad and elsewhere.

Making mud-resist paste by kneading mud and other substances with the feet. Bagru, Rajasthan, India, 1994.

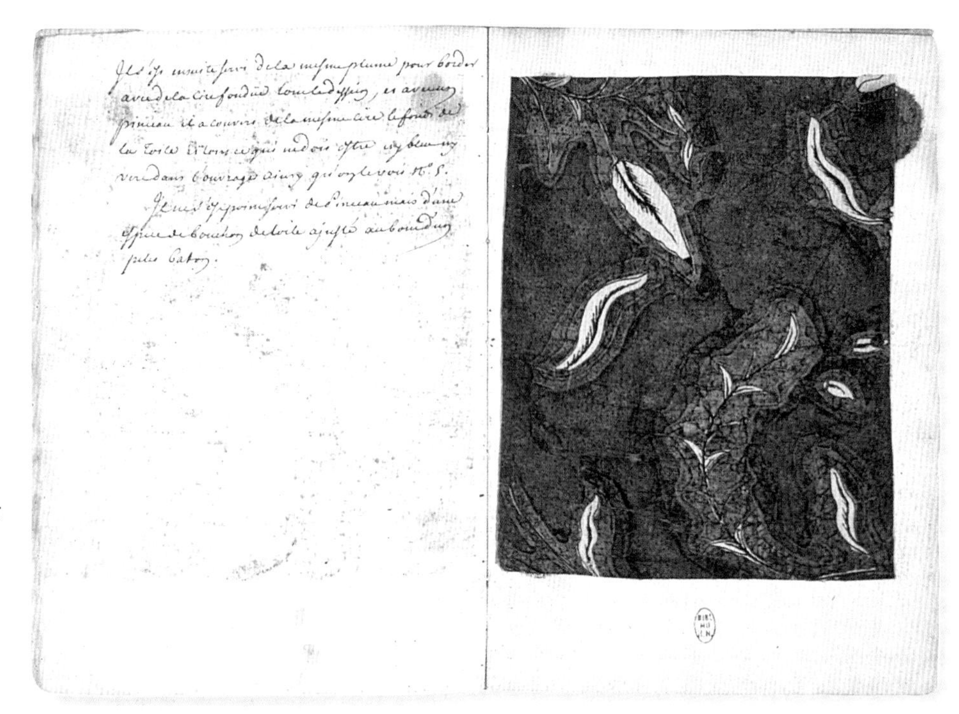

Il s'est ensuite servi de la mesme plume pour border avec de la cire fondue tout le dessein, et avec un pinceau il a couvert de la mesme cire le fond de la toile et tout ce qui ne doit estre ny bleu ny vert dans l'ouvrage ainsy qu'on le voit N.° 5.

Il ne s'est point servi de pinceau mais d'une espece de bouchon de toile ajusté au bout d'un petit baton.

Sample in the Beaulieu manuscript showing the procedure for achieving indigo-blue patterns on a *kalamkari*, painted cotton cloth, in Pondicherry in the 1730s. Hot wax (stained blue as re-used) has been applied to all areas not intended to be indigo blue. (See also p.50.)

Il a plié la toile en plis de 4 pouces ou environ et apres l'avoir un peu pressée il l'a trempée plusieurs fois et ensuite dans une jarre pleine de liqueur propre a teindre en bleu, et ayant etendu la toile il a appliqué de cette liqueur sur les endroits qu'il a jugé n'en avoir pas assez pris, ensuite il a etendu la toile a l'ombre pour la faire secher.

Il a enlevé la cire de la toile en la plongeant plusieurs fois dans l'eau chaude et changeant l'eau de temps en temps

La cire estant detachée il a donné a la toile 3 lessives avec l'eau de la fiente de cabri et l'exposant tous les jours au soleil en l'arrosant comme il a esté dit dans l'article de la 3.e Preparation Il l'a ensuite fait secher et on en a coupé le 6.e morceau

Sample in the Beaulieu manuscript showing the areas dyed indigo blue after wax-resist has been removed.

The finest coloured cottons, *kalamkaris*, which took both East and West by storm and whose trade reached a peak in the seventeenth and eighteenth centuries, came from the Coromandel coast of southern India. (Today production centres thrive again in Andhra Pradesh at Masulipatam and Sri Kalahasti,[31] although they use a minimum of indigo blue as it is still an awkward customer.) The word *kalamkari* derives from the Persian *kalam*, meaning 'pen', for the best cloths were entirely hand painted. Where work was produced under the patronage of local temples and pictorial cloths followed a religious colour code, indigo blue was associated with the god Krishna.[32] *Kalamkaris* destined for the home market had dark grounds which avoided the extensive application of wax-resist, but those made to order for export by the East India Companies to suit European preferences had pale grounds. This gave craftsmen yet another headache, for it meant that as much as 90 per cent of the cloth had to be covered in resist in order to keep the ground undyed and obtain small areas of blue.

De Beaulieu, a French naval officer, made a detailed record of the painstaking *kalamkari* process in Pondicherry in the 1730s.[33] His manuscript contains annotated samples illustrating every stage in the production of a multi-coloured painted cotton. Complicated treatments of the cloth with dung and mordants preceded the painting of mordant dyes. Then everything not intended to be blue or green had to be covered in beeswax before the cloth went off to specialists to be dyed in indigo. All the wax then had to be boiled off before further colours were created, which involved yet more bleaching in dung and applications of mordants and dyes. Yellows were painted over blue areas to make greens, but these eventually faded back to blue. De Beaulieu's record, and other similar European eyewitness accounts, helped European calico printers to develop their own colourfast patterning methods to rival those of India.[34]

Europe was crazy about Indian chintzes, first used mainly for bed-hangings but later for furnishing and clothing generally. They were a great novelty and must have looked positively fresh and sparkling to European eyes so long used to woven textiles, mostly of wool. Various official measures taken to ban their import in the late seventeenth and early eighteenth centuries were, like the ban on Indian indigo, bound to fail, as such rearguard measures usually do.[35] It was inevitable, too, that before long European textile manufacturers would start manufacturing their own versions to suit the public taste. This led to a love-affair with blue-and-white resist-printed textiles in central Europe, and a long search in western Europe for direct ways of printing indigo alongside madder to avoid the need to apply resists before the indigo-dyeing stage.[36]

Dyeing a mud-resist and mordant-printed cloth in indigo. Bagru, Rajasthan, India, 1994.

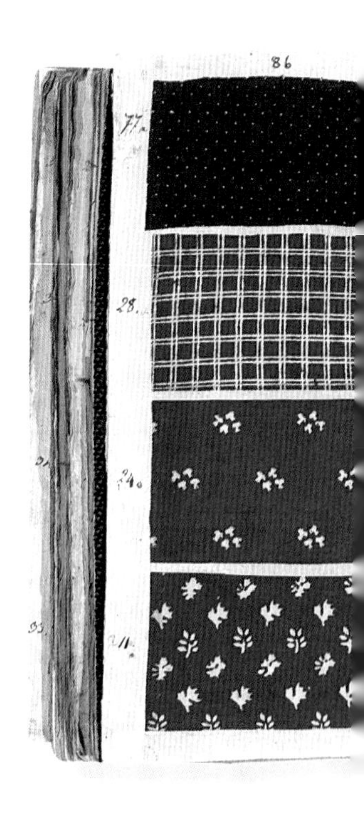

Sample book showing resist-printed fabrics dyed in indigo. Alsace, France, eighteenth or nineteenth century.

Europe's experiments to create her own chintzes, or *indiennes*, date to the mid-seventeenth century, when printers sought to progress from block-printing with oil-bound pigments that were not washable to printing with mordants and resists that could be used with colourfast madder and indigo dyes. This coincided with the increased availability of both indigo and cotton, which gave Holland a head start as she had no woad industry to protect. Resists like wax would have melted in the warm woad vats but now they could be painted on by hand in the Indian manner and be immersed in the cooler woad/indigo or indigo vat, which was also suited to cotton. In the first half of the eighteenth century, when the cold indigo ferrous sulphate vat was also being introduced, metal and wooden printing blocks for applying hot wax and paste came on the scene.[37] By the middle of the century indigo resist-printed cloth was popular in Europe and America.[38] The combination of block-printing a resist and dyeing with indigo, sometimes known as *imprimerie de porcelaine*, allowed the blue-and-white colour scheme so popular in ceramics to be echoed in clothing and furnishings. This system still exists in a few parts of central Europe (see below). However, after much trial and error, western European printers at last discovered colourfast ways to paint and print *directly* with indigo, thus dispensing with resists. Of this innovation one expert has commented that 'this was perhaps the most decisive advance in the early history of textile printing'.[39]

Direct printing with indigo – 'Pencil' and 'China' Blue

If indigo was to be applied directly onto cloth a way had to be found to prevent indigo white from oxidizing and changing to blue before it reached the fibres. Two distinct methods came to be used, both thought to have been developed in England as they were often referred to as 'English Blue' in continental Europe. The first, known as 'Pencil Blue', was introduced in the 1730s. It involved adding orpiment (an arsenic trisulphide) to a ferrous-sulphate indigo vat thickened with Gum Senegal. The orpiment retained the dye in its undeveloped state just long enough during its passage from dye pot to cloth for it to be applied by brush, or 'pencil', before it oxidized.[40] This enabled the calico-printing industry for the first time to paint directly with indigo, thereby creating positive blue designs. But 'Pencil Blue' had to be applied quickly in small amounts and was not entirely satisfactory when block-printed.[41]

Resist-printed cloth, stretched or star frame, being winched out of indigo vat. The characteristic gre colour of indigo before it has oxi can be clearly seen. Workshop o Josef Koó, Steinberg, Austria, 19

The other method, 'China Blue', which was secretly developed in England by the 1740s, was based on a different chemical principle from 'Pencil Blue', since it involved printing indigo in its undissolved state, followed by solution and reduction of the dye *after* printing. In practice this meant printing a paste made of finely ground indigo and a thickening

agent. The blue subsequently developed during alternate immersions of the printed cloth in baths of lime (to dissolve the indigo) and ferrous sulphate (to reduce it).[42] At this stage the French ban on *indiennes* was theoretically still in force, but the Oberkampf factory at Jouy adopted the 'China Blue' method themselves. It produced well-defined, even blues for monochrome engraved copperplate printing and for speedier roller-printing, but the process damaged madder dyes.[43] 'Pencil Blue', on the other hand, enabled touches of blue and green to be added to polychrome madder-based 'full chintzes'. Both methods remained in use in the nineteenth century, despite the introduction of Prussian Blue which was easier to use but harsher-coloured. One related mystery is the existence of many splendid 'American blue prints', whose provenance is unknown.[44]

By 1890 direct blue printing with indigo was used less and less, although improved recipes had been discovered. One was a 'glucose process' (still used today), which involves applying glucose prior to printing with an alkaline print paste containing indigo.[45] Another is the 'hydrosulphite process', where all the chemical reagents are contained in the print paste, but the printed cloth then has to be steamed. The latter system is more expensive but has many advantages, one of which is that it enables indigo to be printed alongside other modern vat dyes.[46]

'Blue-printing' (*Blaudruck*) of central Europe

The so-called 'blue-print', *Blaudruck*, once produced in small towns and villages throughout most of central Europe, was initially worn by the bourgeoisie in the early eighteenth century. By the early nineteenth century, according to Josef Vydra's classic work on the subject,[47] it was in vogue at all levels of society. Its use remained widespread in popular costume in rural areas until the Second World War and it is still very much embedded in folk tradition. One excellent display of *Blaudruck* costumes and printing equipment is in a former *Blaudruck* (or *kékfestö*) factory at Pápa in Hungary. In a few places, including Moravia and Slovakia (where it is called *modrotlac*) in former Czechoslovakia, *Blaudruck* is even still block-printed in the traditional way by hand[48] and dyed in zinc-lime or ferrous sulphate vats. Other places produce imitations by modern methods, using harsher-coloured indanthren blue dye.

These 'blue-prints' are in fact resist-printed and dyed in indigo by a technique long since vanished in western Europe. The basic ingredients of the starch-based resist paste are physical agents such as kaolin, gum, tallow and lard, plus chemical agents such as salts of copper, aluminium and zinc.[49] If chromium salts are added to the paste the resulting designs appear in yellow, orange or green instead of white. The ingredients are cooked to a sticky 'pap', spread across a printer's tray and then transferred onto printing blocks. These were traditionally carved by specialists out of pear or other fruit wood, often combined with metal strips and brass pins, and every blue-printer would have hundreds of different designs. The greatest skill lies in transferring just the right amount of 'pap' onto the block, in order to print repeat patterns rhythmically onto cloth stretched out on the printing table. After fabric is dyed in indigo it is 'soured' in a weak bath of sulphuric acid, which removes most of the paste from the cloth.

An ingenious machine called a 'Perrotine', which mechanized block-printing, was invented in Rouen in 1835,[50] and it remained popular with the blue-printers of central Europe until at least the 1940s. Another invention was the outsize box mangle (see p.137) used to give a lustre to indigo-patterned cloth, especially for festive skirts. These were permanently pleated, the pleats apparently being set with steaming loaves of bread.[51] Today *Blaudruck* cloth is calendered between hot aluminium rollers.

Indigo-dyed cloth patterned by means of a chemical discharge to give a resist-printed appearance. Manufactured at Mycocks of Manchester until 1992 for export to South Africa, where it was known as *chwe-chwe* cloth.

Indigo discharge printing

One chemical method used with indigo is the 'discharge' technique, also known as 'pattern-bleaching'. In contrast with the resist methods, the cloth is first dyed in indigo and then a bleaching agent is applied to remove colour from parts of the cloth already dyed. This means there is no paste to wash off, although the process is still lengthy. Discharged areas may subsequently be overprinted with other colours if desired. Discharge printing has generally been considered inferior to mechanical resist-printing but it was cheaper and the results can sometimes be indistinguishable. Chemical bleaching began in the late eighteenth century in Europe, and a workable discharge for indigo was discovered early the following century.[52] It was not until the 1870s, however, that a French chemist came up with a better method involving sodium chlorate, potassium ferricyanide and citric acid.[53] Further improvements came along in the first decades of the twentieth century.[54]

In the second half of the nineteenth century some European factories hit on the idea of using indigo discharge to imitate genuine resist-printed *Blaudruck*. Such cloth was intended for export to South Africa, to be sold

to German and Dutch settlers who were accustomed to the look of *Blaudruck*. English missionaries subsequently made use of the cloth to cover up the bare-breasted indigenous Xhosa and Zulus, who then absorbed the fabric into their own culture. Such cloth was being produced by Mycocks of Manchester right up until 1992, and similar versions are still produced locally in South Africa using British methods and equipment.[55]

In England, the craftsman William Morris was doggedly determined, in the 1870s and 1880s, to perfect the art of hand-block printing with indigo discharge as his trials with 'Pencil Blue' were unsatisfying. His obsession with indigo, revealed in letters to his friends,[56] made him the butt of many jokes; as his hands and arms were usually stained blue and he wore classic blue workclothes, his friends nicknamed him 'Blue Topsy'. It actually took Morris eight years to fully master the indigo discharge process, and he meticulously recorded his recipes in his Merton Abbey dye book.[57] By varying the amount of colour discharged, and overprinting with reds and yellows, Morris achieved an extensive range of subtle shades. So excited was he by the indigo discharge technique that he used it for nearly all the popular patterns registered by his company between May 1882 and September 1885, and his indigo dye vats remained in use until the firm closed in 1940.[58]

Max Beerbohm cartoon entitled 'Topsy and Ned Jones settled on the settle in Red Lion Square'. William Morris ('Topsy') is caricatured wearing his characteristic indigo-dyed working outfit.

Indigo discharge print by William Morris called 'Eyebright', 1883.

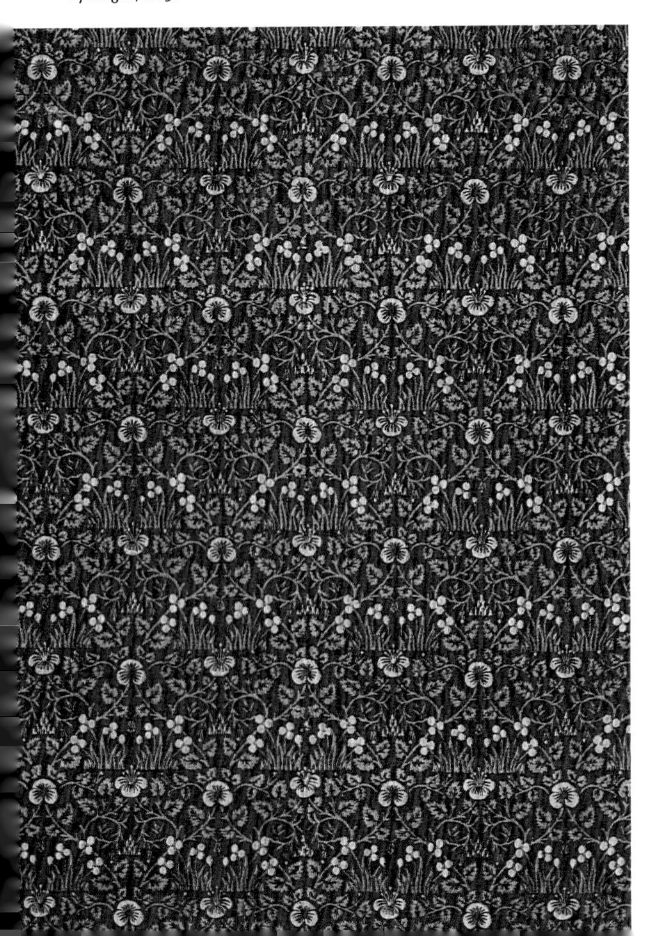

Clamp-resist

A further unusual but effective resist method known as 'clamp-resist' was invented in China in the fourth–fifth centuries, where it was known as *jia zie* (board-jamming).[59] The technique spread to Japan, India, central Asia and southern Europe, but died out in recent centuries.[60] The oldest examples are indigo-dyed. The system involved shielding parts of fabric not intended to be dyed by sandwiching them tightly between pairs of hard wooden blocks which had matching designs carved on one side in relief. On the other side of the blocks were cut deep channels and holes through which dyes were poured to penetrate designated parts of the pattern. Modern textile artists have returned to the clamping system in innovative ways.

Soft-edge patterning

Indigo, again because of its unique chemistry, is the dye *par excellence* for all the soft-edge resist techniques. These cause the dye to penetrate the fabric in varying degrees, and subsequent oxidation to be uneven. The unpredictable randomness of the results can be a revelation. Soft-edged methods can be broadly divided into those applied to yarn *before* it is woven, known as ikat, and those applied afterwards by manipulating cloth and binding or stitching it in various ways.

Silk and cotton ikat textile from the Cevennes, France, nineteenth century.

Ikat

Whatever the origins of ikat, the technique has a long history, as it appears in archaeological textiles from the Near to the Far East, and in cave paintings of the seventh century AD at Ajanta in India. It was known in India, Indonesia, Japan, Central and South America, Africa and central Asia.[61] Fine ikats were also made in many parts of Europe, due probably to Islamic influences, the *chiné* of Lyons being the most exotic.

Picasso was intrigued by ikat. 'If I weren't so busy with pottery,' he once said, 'I might perhaps try my hand at ikat; it is a remarkable art form, exciting and transcendental.'[62] No doubt the element of surprise inherent in the craft of making an ikat attracted him, for the finished result, no matter how carefully pre-planned, remains unpredictable. This is related to the characteristic blurring of the pattern edges as they blend into the background. The process involves tightly binding (or clamping) bundles of yarn before dyeing to prevent dye from penetrating pre-arranged patterns, and removing all binding threads before weaving. The tying, untying, re-tying (to get colour variations) and re-ordering of the bundles, and then the arranging of the dyed threads on the loom, can get extremely complicated.[63] The weaving process is quite straightforward for single warp or weft ikat, but aligning the warps and wefts of compound and double ikats can be much more challenging. Famous double ikats are the *patola* of Gujarat, India, and the *geringsing* of Tenganan in eastern Bali.

Ikat has long been practised by women throughout the Indonesian archipelago and in the Philippines, but it is fast dying out. The arrangements of reds and blues distinguished communities and were closely associated with ancient traditions and religions.[64] Most Indonesian ikats were dyed with morinda red and indigo, often combined to create browns or 'black'. To achieve really strong colours, the dyeing of tied yarns could literally take years, and this is still intrinsic to the creation of certain ceremonial textiles. Those of cotton are usually warp ikats, a formal example of which is the geometric blue *ulos* of north Sumatra, whereas silk was traditionally weft patterned (by Muslim weavers).

Indigo-dyed ikat is found in many other parts of Asia. North-east Thailand, for example, is still known for its cotton *matmi/mudmee* ('patterned threads'), dyed in natural indigo.[65] Ikat, like other soft-edged methods described below, is particularly suited to the aesthetic sensibility of the Japanese, who are famous for the subtle and delicate designs created on a wide variety of fibres in both weft and double ikat, *kasuri*, still worn by some people today. From the eighteenth century ikat fabrics, *kon kasuri*, typically indigo blue and white, in both geometric and pictorial designs, began to feature widely in all types of clothing and household items.[66]

In central Asia, apart from the early indigo and yellow Yemeni ikats which were renowned far and wide in the early Middle Ages,[67] certain cities have been known for their silk ikat textiles. In the 1970s the last indigo dyers of Aleppo were dyeing ikat yarns tied up by a specialist called a *rabbat*, 'tie-man', to be woven into sumptuous silk coats.[68] Most gorgeous of all are the silk ikat garments first produced in Turkestan's Bukhara and Samarkand in the nineteenth century, the dyeing of whose colours was, as noted, divided between those who dyed the 'hot colours' and those who dyed the indigo blues.[69]

[top]
Tying weft yarns to reserve patterns before dyeing in indigo, removing the bindings and weaving the yarns. Ba Saang Sa, village weavers' project, near Nong Khai, north-east Thailand, 1991.

[above]
Weaving indigo-dyed ikat yarn. Bukay, Ecuador, 1988.

In West Africa, where the use of both indigo and ikat almost certainly arrived with Islam, and the two were always combined, ikat was woven by men in two main centres.[70] In the Ivory Coast Dyula and Baule dyers were famous for it, and in south-western Nigeria the technique was practised by several ethnic groups, including the Yoruba. Indigo ikat has also been woven by northern Nigerian Hausa, and by the Ewe of south-eastern Ghana. On the other side of Africa warp ikat was until recently woven in Madagascar, whose exotic textile arts were strongly influenced by early trade links with Asia.[71]

The other region of the world most noted for indigo and white ikat is Central and South America, where warp, weft and compound ikats have all been widely woven on both backstrap and treadle looms for use as skirts and shawls. The technique, known as *jaspeado* ('speckled'), was used in Peru, Mexico, El Salvador, Ecuador and Guatemala (where indigo-dyed ikat is still made in Salcajá).[72] Down the Pacific coast the Chilean Mapuche people made exceptional ponchos and horse blankets for ceremonial use in strong geometric designs of dark indigo blue (see p.190).

Central Asian (Turkestani) silk ikat coat, 1860s. The blues were dyed by Jews, the 'hot' colours by Tadjiks.

Cotton ikat *tzute* (head covering). From San Cristobal Totonicapan, Guatemala, twentieth century.

Shibori – *plangi* and *tritik*

Soft-edge resist methods of handling cloth, as opposed to yarn, are legion. Often combined, they are achieved by various modes of manipulation, which are almost sculptural, and include gathering, tying, scrumpling, binding, stitching, tucking and folding. In Japan they are known collectively as *shibori*, i.e. 'squeezed'.[73] This useful term has been widely adopted to embrace two different methods. One is *plangi*, or 'tie and dye', which involves binding up cloth or wrapping it round small objects prior to dyeing. The other method is *tritik*, or stitch-resist, where cloth is sewn, and usually gathered, before being dyed. *Tritik* used always to be done by hand, but today the sewing machine is often used to speed things up. There is always an exciting element of surprise for the dyer when ties are unpicked and cloth is folded out. Sometimes unrepeatable marble-like explosions are created, and no dye is more naturally suited to *shibori* than indigo.

Shibori evolved in ancient times, as is evident in archaeological textiles from places as far apart as Peru and China. It is common in much of Asia, but is most highly developed in Japan, where it was known by the eighth century but blossomed in the Edo period (1600–1868). For the Japanese an indigo-dyed *shibori* textile both celebrates the quality of the dye and enhances the natural texture of the cloth itself. Their dyers have invented a quite breathtaking variety of designs based on binding, stitching, folding and pole-wrapping, and they have given the finished results such aptly poetic names as 'white shadow', 'spider web' and 'angel wings'.[74]

West African dyers, too, have long been known for indigo-dyed *shibori*, the art having perhaps reached them from Asia.[75] An unusual indigo tie-dyed coif from the Tellem caves in Mali (see p.26) shows a familiarity with these techniques nearly a thousand years ago. A classic West African colour combination of indigo with kola nut brown is sadly dying out today.[76] West African dyers have excelled at both *plangi* and *tritik* techniques, and

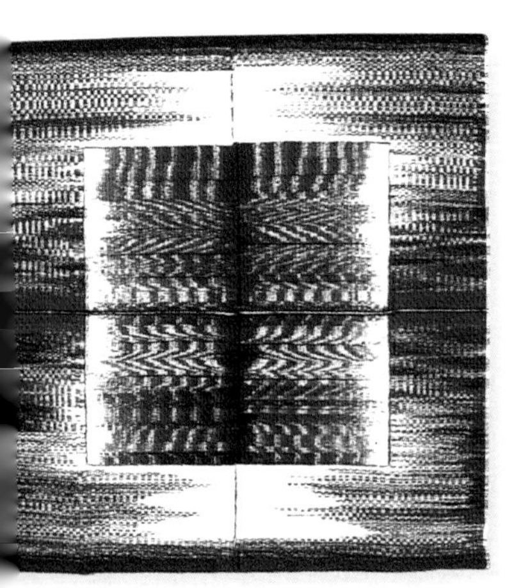

igo water, Creber's Pool, River Tavy' x 67 cm). Indigo ikat tapestry hanging en by Bobbie Cox, Devon, 1993.

Plangi (tie-dyed) and *tritik* (stitch-resist) cloth from Nigeria, with raffia, seeds and stones used to create the patterns. Collected 1960s.

both remain popular but are nowadays produced in every colour under the sun.[77] The most astonishing *tritik* work of all was created for ceremonial use by Soninké and Wolof people along the Senegal river. They embroidered fabric in intricate detail before dyeing it in indigo. When the embroidery threads were removed after dyeing, patterns appeared like embroidery ghosts floating on a backdrop of midnight sky.[78] Such *plangi* techniques as plaiting cloth into ropes before dyeing are still practised in St Louis today, using modern European damask cut into strips, a legacy from the days when narrow strip cloth was standard currency in the region. On both sides of the Senegal river and elsewhere stitch-resist is still used to pattern indigo-dyed cloth. Stitching, or tying around stones, grains or sticks, was usually done with raffia, but today cotton thread has largely replaced it.

In Nigeria's Yorubaland tied and sewn cloth is called *adire oniko*, and the designs created are given such names as 'scarification and fingers' and 'little moons'. The symbolism of resist reflects the complexity of the societies concerned. For example, the dramatic resist-dyed indigo cloths called

Samples of embroidered resist, before and after dyeing, and after removing the embroidered threads. St Louis, Senegal, acquired in the 1930s.

ukara, made to order in Nigerian Igboland for ceremonial use by the Ekpe Leopard society further south, are full of powerful symbols such as snakes and swastikas, as well as leopards.[79] These in turn probably influenced the *ndop* of the Cameroon grasslands.

Many fine resist techniques are crudely imitated elsewhere for today's 'ethnic' mass market, as produced, for example, by cheap Chinese labour under Japanese direction. Japan, however, can boast such superlative textile artists as Hiroyuki Shindo, for whom indigo's subtle shades and blended edges create visual poetry.[80]

Installation by Hiroyuki Shindo of indigo-patterned woven hangings and dip-dyed spheres. Gallery Space 21, Tokyo, 1998.

Patterns woven and applied

In addition to all the ways of treating yarn or cloth to create patterns, plain indigo-dyed yarn or cloth is also used in a wide variety of decorative ways.

Weavings limited to blue and white are a speciality of parts of rural West Africa. The weave is usually structured to create bold abstract patterns of stripes and rectangles and relies on contrasting warp or weft threads, simple weft-faced additions and the effect of joining narrow strips together. In the past this usually meant indigo-coloured yarns of every shade contrasting with the natural colour of handspun cotton or other fibres like raffia.[81] The resulting patterns bear such names as 'land tortoise', 'guinea fowl feathers' or 'the eyes which mystify'.[82]

Indigo-blue threads have also figured extensively in embroidery, and again the effect of blue set against a white or off-white background has a compelling simplicity that has been widely appreciated. In Europe the shaded blues of seventeenth-century crewel-work furnishings spring to mind. Other places noted for blue-and-white embroideries were the Moroccan city of Fez and rural provinces of southern China, where embroideries of dark blue silk were a fundamental part of domestic furnishings and clothing until production ceased by 1950. Designs chosen, some dating back to the Tang dynasty (AD 618–906), were believed to confer good luck and protection on married couples and their children.[83]

Coloured embroidery or appliqué work required the darkest possible colourfast background, and indigo-dyed cloth provided just that, avoiding the need for cloth dyed in corrosive black mordant dyes, notably iron. When factory-produced fabrics such as black sateen came on the scene they took over from indigo in many places, but their quality is quite lifeless by comparison. Before this lamentable change-over took place, the Arab world produced much exquisite embroidery on indigo-dyed cotton or linen dresses and coats, famously in Palestine and Syria.[84] The quality of such clothing was captured by Holman Hunt in his painting *The Shadow of Death*.[85] He was clearly also fascinated by the sheen of polished indigo, depicted in two versions of his *The Afterglow in Egypt*.[86] Such lustrous indigo cloth was also used for many exotic embroidered dresses of Yemen and the Asir region of Saudi Arabia.

Once again it is among the ethnic minority groups spanning southwest China, northern Thailand, Burma, Laos and Vietnam that glazed and burnished indigo-dyed cloth is still widely used. Men and boys in remote areas will wear it unadorned, while women use it for spectacular garments embellished with all kinds of decorative technique.[87] These include fine cross-stitch, couched work, satin stitch, folded silk, strips of tin, batik bindings, appliqué panels and floating embroidery featuring, for instance,

Cotton robe from Liberia, the bold checked pattern formed by being woven in blocks of indigo and white in 13 cm strips (height 112 cm).

Palestinian dress of indigo-dyed handwoven linen with silk embroidery and appliqué work. Yatta, southern Hebron hills, 1930s or earlier.

'Travellers passing a shrine in the mist', coloured woodblock, Hiroshige 1831–41. The travellers are wearing patterned indigo-dyed clothing, the mounted person appears to be wearing double ikat cloth. The traveller on the right has a *furoshiki* on a pole over his shoulder.

vivid butterflies that appear poised to fly off their indigo ground. When embroidered cloth grows grubby and tired, it is often re-dipped in an indigo dye vat and acquires a new character all of its own.

Among the simplest but most special indigo cloths are those decorated with quilting by a technique known in Japan as *sashiko* ('stitching'), a good example of the way Japanese culture successfully marries the functional and the aesthetic.[88] *Sashiko* quilting developed out of a need to reinforce hemp cloth at points susceptible to wear and tear, recycle old cloth into new garments and join layers of fabric together for warmth. The large square wrapping cloths, *furoshiki*, which were used in Japan as universal carry-alls and appear in many Japanese prints as part of a traveller's baggage, were both beautiful and practical three-dimensional objects. In the past many garments such as farmers' jackets and the *kogin* garments of northern Honshu made practical use of *sashiko* stitching. Today, however, *sashiko* is usually employed for purely decorative reasons, practised as a sophisticated technique in the West as well as in Japan.

(far left)
Miao women in jackets of indigo-dyed, handwoven, diamond twill cloth with richly embroidered sleeves. Langde, Guizhou, south-west China, 1993.

Large square *furoshiki* used in Japan as a universal carry-all, the corners strengthened with *sashiko* quilting. Sets of *furoshiki*, usually indigo blue, played an important role in rural marriage ceremonies, and *furoshiki* gifts are common today.

Boys and girls wearing festive indigo-dyed costumes dancing on a hillside near Taijiang, Guizhou, south-west China, 1993.

The beauty of unadorned indigo-dyed cloth, whether faded and well loved or with a sheen to show off, should also be mentioned.The appeal of a new shiny indigo-dyed turban or, for a woman, a dramatic face-mask has already been described (see p.133). But I shall close this celebration of indigo's inimitable qualities with an unforgettable personal memory from south-west China. This is the sight of girls and boys, all wearing indigo outfits, dancing in a ring on a remote mountain terrace, the girls' swirling skirts made from glazed cloth hand-pleated as finely as the gills concealed beneath the cap of a field mushroom.

This vegetable dye is in demand,
from the imperial robe to the peasant's stocking,
and forms alike the delicate white of the muslin dress,
and the dark blue of the gardener's apron.
(HENRY PHILLIPS, 1822)

'For Richer, *for* Poorer'

Textiles Prestigious and Popular

The status of indigo-dyed textiles across the world has turned almost full circle over the centuries from its earliest history to the present day. Initially, because of its rarity, indigo featured only in luxury textiles, next in prestige to those dyed with shellfish purple. When indigo dyestuff became progressively available worldwide, it passed into common usage and many centuries later into blue denim. Today, though still easy to come by, indigo's special qualities have been seized upon by fashion producers to give it once more a new and pricey prestige. China provides a prime example of this circular story. In early imperial days blue was reserved for princes and nobles, but by the time of the communist revolution its use was widespread amongst the poor. In 1929 an official ruling that everyone should wear the kind of modest clothing already worn by the peasantry gave birth to the uniform indigo-blue 'Mao suit', which became a symbol of the communist movement's 'popular' base. Since this sartorial restraint was lifted, Chinese rural style workwear is resurfacing as fashionable clothing in high streets elsewhere.

Other factors have variously affected the status of indigo-dyed clothing in different parts of the world. There are countries where the strange behaviour of the dye has invested it with spiritual significance, setting it apart from other dyestuffs and resulting in its special use for special purposes. For some people indigo-dyed cloth acquires a symbolic prestige

Ceremonial garment formerly worn for certain religious rituals in southern India around the turn of the nineteenth/twentieth century. Embroidered on an indigo-dyed ground. Labelled 'devil dancing trousers' by the collector.

Bands of indigo-dyed cotton inserted above the hem of a Jordanian bedouin wedding dress otherwise made of black sateen. The indigo-blue cloth was thought to have protective powers. Early twentieth century.

from the extra depth and reflective shininess bestowed upon it by hard-working dyers; for others it is the beauty of the colour in its many shades, as in oriental rugs, that arouses admiration. The Japanese have separate names for the various blue shades resulting from different numbers of immersions in the indigo vat. Likewise eighteenth-century European dyers classified thirteen separate shades of indigo in common use. Beginning with the lightest, these were 'Milk-blue, pearl-blue, pale-blue, flat-blue, middling-blue, sky-blue, queen's-blue, turkish-blue, watchet-blue, garter-blue, mazareen-blue, deep-blue, and infernal- or navy-blue'.[1] 'Queen's' (or 'king's') blue was also called 'royal blue', a description that like 'navy blue' has left its indigo roots behind and passed into common colour terminology today.

Even in countries where depth of colour is considered a special virtue, dark indigo was often regarded as superior to black. In Morocco, for example, certain Berber shawls striped indigo and black from other sources were valued proportionally to the ratio of indigo to black.[2] In some countries the dangerous but protective powers ascribed to indigo have led to its insertion in costumes as a kind of sartorial homoeopathic repellent. For instance, in parts of the Levant where black sateen has taken over from indigo, horizontal bands of indigo-blue cloth are still to be found inserted near the hems of wedding dresses to ward off the evil eye. In other countries indigo is a symbol of national identity and is therefore worn on state occasions as well as at weddings and other ceremonies.

In the more detailed examinations of the subject that follow, the choice of examples will focus especially on indigo's role in ceremonial and ritual textiles, since the elaborate symbolism involved has been progressively losing its potency and it risks being forgotten.

Ritual and ceremonial textiles and beliefs

There is no universally symbolic colour that marks life's important rites of passage. However, for many diverse ethnic groups indigo-dyed textiles have been significant for reasons which went beyond colour *per se* and the widespread availability of the dye. The association of indigo with death has been particularly apparent, and in some societies as disparate as the Dogon of Mali and the Sahu people of the Indonesian Moluccas special indigo-dyed textiles have been worn to mark stages in the agricultural cycle which parallels the human progression through life and beyond. Thus the Sahu wear indigo-dyed sarongs at the close of the harvest festival and red and yellow ones (colours of the new cycle) the following day.[3] As death and separation are allied to the regeneration of life and, by extension, to fertility, there is no contradiction for those who have linked the living fertile dye with birth, marriage and death equally. In Java, for example, blue shades are considered female but also relate to age and death,[4] and in other parts of Southeast Asia red is often associated with masculinity and the right side, while indigo 'black' stands for femininity and the left.[5] The distinctive smell of an indigo dye vat has itself a pungency reminiscent both of female odours and death and putrefaction. These multiple associations, absorbed into myth and superstition, have gone hand in hand with the sense of awe inspired by the magic behaviour of the dye. But which was the chicken and which was the egg? Even today, despite much de-mythologizing, indigo still retains for many people at least some of its accumulated aura. This is particularly the case where tradition still demands that, whatever daily clothes are made of, cloth for ceremonial occasions should be made by hand from indigenous materials. There are several possible reasons for the original use of indigo in such textiles, but by becoming customary it may have acquired a special significance, much as the wearing of dark clothes became so-to-speak sanctified by force of habit.

Sarong of the Manggarai people of western Flores, Indonesia. Indigo-blue with geometrical motifs woven in a supplementary-weft technique.

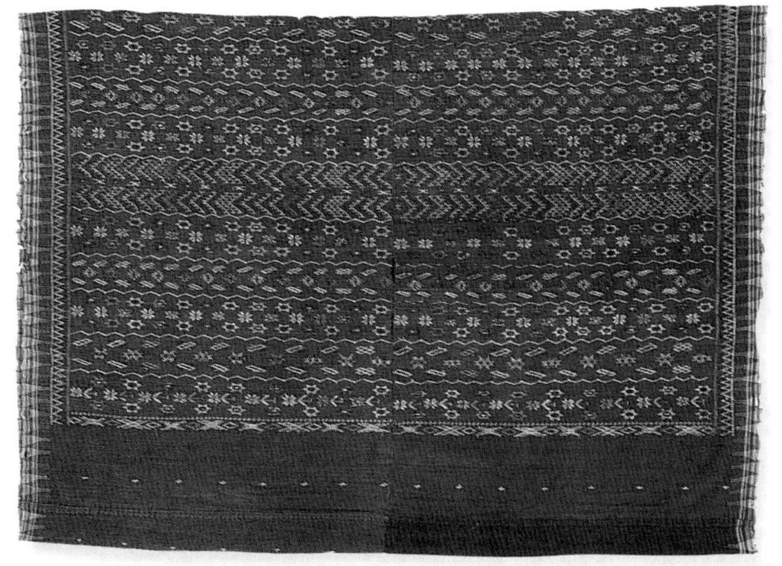

Indigo is of course by no means the only dye colour to have accumulated symbolism. Red, representing blood, has often been the symbolical colour of marriage, but in ancient Peru it stood for death. In some places in Indonesia indigo is overdyed with morinda red and the indigo component is deliberately 'hidden'[6] One example is the famous double-ikat ceremonial cloth, *geringsing*, as has already been described. In such cases the indigo/red combination produces a characteristic rich brown considered by some to resemble dried human blood.[7]

Double ikat cotton *geringsing*. The dyeing of these sacred cloths with morinda red and indigo (the latter dyed in a neighbouring village) could take many years.

It is impossible to cover all the conflicting aspects of indigo's use in ceremonial textiles but its unique chemistry provides once again the key. The transformations involved in the vat dyeing process, akin to alchemy, appeared to mirror the mysterious changes in the process of human life itself. And the sudden 'death' of an indigo vat has been linked, as we have seen, with nearby human death or disease. Such mysteries have been both revered and feared, and have slotted into the intricate pattern of cultural values. Surface changes to indigo-dyed cloth, such as making it shiny, have been regarded as imparting to the wearer additional spiritual energy. The many dippings required to make indigo-dyed cloth a dark blue-black colour linked it with wealth and religious prominence.

Textiles made specifically for ritual and ceremonial purposes have for generations played a central role in many societies, but owing to the speed of social change and cultural homogenization many of the associated beliefs and customs are vanishing fast. The discouragement of 'pagan' rites by over-zealous missionaries has also hastened the demise of a rich diversity. It is certainly true that mass tourism has created a fascination with 'strange' products, but for the most part we are now reduced to peering at genuine ceremonial textiles through museum glass, appreciating their beauty but unable to engage with their cultural potency. For enlightenment we must rely on such studies as Gittinger's *Splendid Symbols* or Maxwell's *Textiles of Southeast Asia*,[8] but Maxwell gloomily concludes that decorative textiles once believed to embody sacred and magical powers will in future 'be used only as anachronistic costume on ceremonial occasions'.[9] Anthropologists spend years seeking to establish an understanding of their full meanings. Sometimes the mysteries can never be fully unravelled, since the function of many sacred textiles is precisely to represent what cannot be expressed in words.

The Near East

In Egypt in the past Coptic and Muslim women followed a custom at funerals of formalized public mourning, which sprang from Pharaonic antecedents. Once indigo became widely available it would have been an obvious choice for funerary textiles and rituals, not only for the reasons already mentioned but perhaps even for its resemblance to the colour of embalmed corpses, made blue by the preservative resins. Coptic women took their indigo rituals to extremes, especially on the death of the head of a household. Not only clothing but even curtains and sheets were taken to the dyer to be immersed in the indigo vat. Poorer women, both Copt and Muslim, would wear their oldest indigo dresses and head-dresses, with applied ornaments removed, or have their other clothing overdyed with indigo. Indigo dye was also smeared on a mourner's face and body, on the walls of the deceased's house and on the funeral drum itself.[10]

In Palestine dark blue or black girdles and caps were worn by women in mourning, and sometimes colourful embroidery was overdyed in indigo, although white was the colour for shrouds.[11] In Oman a piece of polished indigo cloth sometimes covered the corpse beneath the visible white one.[12] In eighteenth-century Syria there was a demand for woollen bier-cloth of indigo-and-white stripes, woven specifically for the purpose in Kashmir.[13] Could this long association between indigo and death explain why orders giving the death sentence for criminals were in the past written on blue paper in Egypt and Syria?[14]

Arab woman in indigo clothing (left) at a graveside with other mourners. Illustration in a thirteenth-century *Maqamat of al-Hariri* manuscript.

Blue songs

The Sanskrit/Arabic word for indigo, *nila*, features in funerary laments both in Egypt and in India. The historical association of indigo with death even lingers unflatteringly in various common Egyptian expletives like *gatak nila* (literally, 'may you be indigo-ed'). Tamil women in India call their laments *nilappaddu* (blue, or dark, songs), and the lamenter *nilamma* (blue woman). In Indonesia, too, songs of lament give expression to indigo's relationship with death and separation, including miscarriage.[15] In a more generalized way, Western 'blues' music plays a similar melancholy role.

Central and Southeast Asia

In central Asia, where turquoise blue has been a common mourning colour, muted indigo-dyed versions of the sumptuous silk ikat garments of Turkestan were worn by wealthy older women and those in mourning. Appearing in the mid-nineteenth century, these were made by a technique known as *adras*, which produced distinctive branching patterns on a sombre indigo ground.[16]

As for Southeast Asia with its renowned textile culture, Maxwell tells us that the making of textiles intended for ritual and ceremony 'often involves women in procedures fraught with uncertainty, calls upon special rituals and magical practices, and involves a degree of secrecy at particular stages'.[17] Many special dyed textiles have had enormous spiritual power and were, and sometimes still are, brought out of storage to mark significant rites of passage. The Iban of Sarawak greatly revere their ceremonial ikat hangings, *pua*, all of which still feature natural indigo in their design.[18] Dark indigo may have served a specific function in those *pua* associated with head-hunting expeditions.[19] In the 1930s Iban widows apparently transformed their brighter-coloured ikat wraps into mourning wear by immersing them in the indigo vat, which produced patterns of light blue on a very dark ground.[20] The Isnai people of the Philippines surrounded a corpse with special warp-ikat blankets whose white patterns on an indigo ground symbolized a man's genealogy and journey through life.[21] Gittinger stresses the importance of the actual dyeing processes for Balinese *geringsing* (see p.180), which can take as much as eight years to complete:

The designs on the textiles may identify certain groups, but apparently do not affect the ritual usage of the textiles as much as other qualities. Their efficacy is more strictly aligned with the quality of the dyes . . . When properly executed, the cloth is considered a protection from illness and harm in general . . . When they are used elsewhere on the island, geringsing *are honoured as the most sacred of all Balinese cloths. They form the head rest at tooth-filing ceremonies, and encircle the bride and groom at weddings. Fragments are used in curing rituals, and entire textiles are hung from temples and cremation towers.*[22]

In Java inscriptions dating from the early ninth century AD refer to ceremonial textiles used as gifts, most of them red or blue.[23] Heringa explores the way cloth colours there have been visual expressions of life's journey. She writes: 'In all

Isnai funeral blanket from the Philippines. In the past the pattern could be 'read' by certain knowledgeable people.

Malay style of folded headcloth (also worn by rulers in peninsular Malaysia) from the Jambi district of central Sumatra, an indigo area. Such headcloths, believed to confer supernatural power on the wearer, are still worn on ceremonial occasions.

overdyed cloths the indigo functions as a means to disguise the other colours. This is in accordance with the myth of origin, the *Manikmaya*, in which indigo is said to be the symbol of the night.'[24] One dark ritual cloth, *irengen*, is used as a shroud and is produced from soga brown overdyed with indigo, which symbolize between them the combination of male and female. Another ritual cloth, *putihan*, is used for life ceremonies and has a white background with dark indigo patterns. The indigo is dyed seven times to produce this cloth, which is never washed as this would weaken its magic power.[25] The blue-and-white combination in textiles was considered generally propitious in both Java and Bali, which is why banners in these colours were hung up in times of community crisis and used in exorcism rituals.[26] In Bali brighter indigo-and-white batik cloths, *kelengan*, resembling Chinese porcelain were worn by a bride on her wedding night; and a symbolic cloth with a striking Islamic cosmological design of white on 'heavenly' blue might also adorn a wedding bed or throne.[27] In central Java a dramatic wedding cloth, *dodot*, typically with a large white diamond on dark indigo ground (with surface gold tracery) symbolized the opposites of earth and water, male and female, life and death. Because of its potency it was reserved for royal use.[28] In many countries what was formerly daily wear is now worn, rather like the kilt in Scotland, only on special occasions. In rural Java, for example, trousers simply batiked in indigo and a wood dye were formerly daily wear but now form part of a bridegroom's wedding attire.

In Sumatra indigo was one of the dyes used on the famous ceremonial 'ship cloths', *palepai*, of Lampung in the south. Their full symbolism is lost to the present generation, but Gittinger, who considers that their colour and design were expressively matched, has suggested that 'the blue ship compositions ... seem to represent an earthly realm, as opposed to the sacred sphere of the red ship *palepai*'.[29] Amongst the Toba Batak of northern Sumatra a local myth relates that an important goddess, the founding matriarch of the Batak people, used an indigo-dyed yarn to escape the underworld and reach the seas, from where she created the earth. She then passed on her vast textile knowledge before retreating to the moon to spin. Special warp-ikat rectangular cloths called *ulos*, which represent a link between the wearer and the spirit world, are still essential wear at ritual events. Indigo-blue *ulos* feature as part of a complex system of gift exchanges at both weddings and funerals, and in the past one of the most important textile gifts was a circular indigo-dyed *ulos* which symbolized eternal life.[30] In Batak society it was often colour rather than pattern that bestowed symbolic significance, white meaning purity, red bravery and indigo-'black' eternity. Each colour represented a deity, and the three

Warp ikat shawl, *selendang*, worn over the shoulder. Savu island, Indonesia.

together signify a unified cosmos. Throughout Southeast Asia the three colours red, white and blue were often combined in textile accessories to bestow talismanic protection on a warrior going off to battle.[31]

In eastern Indonesia the ceremonial significance of dyes varies from island to island. Sometimes red carries the more potent symbolism and sometimes indigo blue/black.[32] On Roti, where textile-lined coffins are ritually tried out prior to death itself,[33] and on neighbouring islands like Savu, indigo is associated with both malevolent spirits and protective powers.[34] Evil spirits can terrify or 'blue' their victims;[35] yet, conversely, indigo's powerful influence can also ward off evil spirits. If a person dies a 'bad death', blood money can be paid to enable the contents of an indigo dye pot to be poured over the victim's grave to prevent the soul from harmful wandering.

On the island of Sumba, where an unlucky death is also called a 'blue death', traditional burial rites are exceptionally elaborate. A corpse could be wrapped in up to a hundred warp-ikat shrouds. In the west of the island Hoskins tells us that 'because of the constant demand for quality indigo-dyed cloth for ritual exchange, the "blue-handed women". . . are highly respected'.[36] In the east of the island the attractive blue-and-white ikat *hinggi kaworu* is man's daily wear, whereas the more prestigious colourful *hinggi kombu* (see p.8), which combines indigo and morinda red dye, is used for gifts of exchange and burial rites.

Man's *hinggi*, cotton wrapper, woven in warp ikat in shades of indigo blue. East Sumba, Indonesia.

Further east in the Solomon Islands a little known category of bark cloth, either dyed all over or patterned with indigo, appears to have had deep-rooted spiritual significance in ceremonies relating to male and female fertility, death and regeneration, and even head-hunting.[37] In the past a bride was presented with a ritual apron of bark cloth dyed dark blue to signify her imminent entry into wifehood, but by the 1920s the use of bark cloth for the apron had given way to blue denim.[38]

China and Japan

From ancient times the Chinese rationalized the world with their beliefs in a balanced hierarchy of symbolic categories. Even the five basic colours to be worn at court played their part in the system of cosmic harmonies. The colours for clothing were allocated to social classes, and those of the emperor himself visually expressed a means of controlling the environment. Blue (though worn by the emperor for certain special ceremonies relating to the sky or to water) was ordained for third-degree princes and nobles. All officials of lesser rank had to wear black.[39] Outside the court the silk costumes worn by the élite of both sexes commonly had an indigo ground or indigo embroidery.[40]

In Japan colours of traditional court dress were similarly prescribed in the ninth century, and dark blues again reserved for those of certain aristocratic ranks. Blue, however, was one of the first colours to be de-restricted, and in the twelfth century became a favourite of the *samurai* warrior class. During the Edo period (1600–1868), when simple indigo-dyed cotton garments became ubiquitous in rural areas, the subdued elegance of blue-and-white resist-dyed costumes made from fine ramie fibres became the prerogative of the highest élite. Purple from a special root dye was reserved for the very highest ranks, but, like shellfish purple in the Near East, was often 'faked' by a combination of indigo and red sappanwood.[41]

West Africa

Indigo has also had a special place in mourning ceremonies and other rituals in West Africa, where beliefs about its origins are, as elsewhere, embedded in religious mythology and folk legends. In parts of Cameroon, Nigeria and Mali indigo has been especially prominent in ceremonial cloths and costumes kept for use at burials and ritual masquerades – masked performances, reserved for males, in which the world of the spirits is invoked. In some cases, as in Southeast Asia, this has helped to preserve the production of handwoven, naturally dyed cloth, but here too such ritual use is fast disappearing.

In central Nigeria the immigrant Ebira people brought with them distinctive masquerading traditions, based on cotton and indigo.[42] The traditional Ebira burial cloth, *itokueta*, differed from everyday cloth by having a weft exclusively of dark indigo-dyed yarn, although it was almost hidden since the warp, distinctively striped in pale and dark indigo and white, predominated. The symbolism of the colours used has been studied by Picton. Indigo has prestigious as well as dangerous associations, but once again such important textiles represent visually what cannot be put into words.[43] Another more prestigious shroud, *itogede* ('banana cloth'),

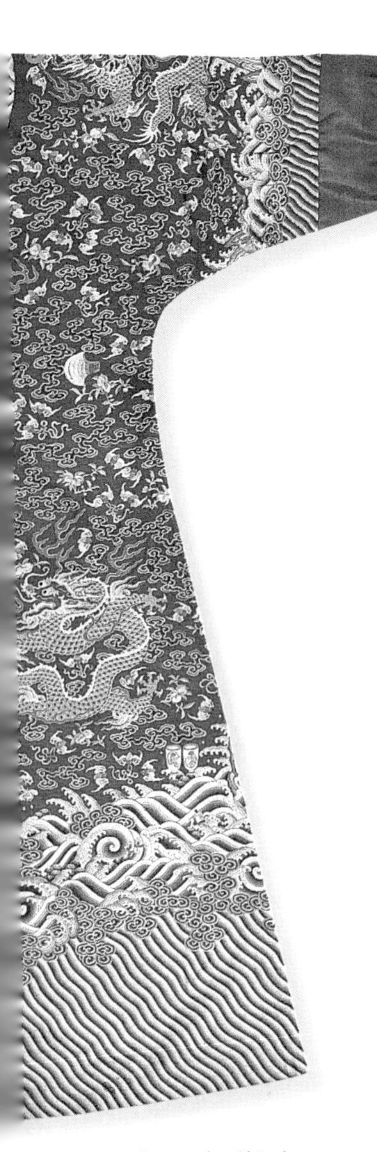
Dragon robe with twelve imperial symbols. Silk tapestry-weave in shades of indigo blue and white. Chinese, nineteenth century.

had an undyed lustrous bast weft with warp stripes of indigo and white.[44] This was reserved for burials of men of high standing. In central Nigeria women of other ethnic groups dyed special head-ties deep indigo blue as part of their own, and their menfolk's, future burial attire.[45]

Among the Nigerian Yoruba, whose use of indigo is renowned, certain prestige indigo or indigo-and-white cloths were reserved for rites and ceremonies. The name of one exceptionally important cloth, the *olowududu*, refers specifically to its very dark indigo colour, *dudu* meaning 'black'.[46] Large robes made from light and dark blue 'guineafowl' check were highly prestigious and often worn at funerals. Even a very poor Yoruba man in the 1960s would try to buy a new indigo-dyed cloth for his wife to mark his appreciation of the gift of a new baby.[47]

Elsewhere in West Africa distinctive uses of ceremonial cloth patterned with indigo include the famous decorated cotton strip cloth, *ndop* (also known as 'Royal', 'Bamoun' or 'Bamileke' cloth), whose complicated symbolism continues to be revered.[48] *Ndop* always features characteristic bold designs on an indigo-blue background. It is used in several regions inhabited by the Cameroon highlanders and, over the border, in the eastern Nigerian district of Gongola. The dyers themselves, who are based in Garoua, refer to *ndop* as 'juju' cloth, reflecting its powerful function as a major feature in funerary rites. Now that its use is no longer strictly controlled by the chief, *ndop* features widely in various ceremonies, notably those connected with funerals, such as masquerades and 'cry-dies'. 'Cry-dies' are occasions of memorial, at which the dancers' skirts and family members' sashes, arm bands or belts are all made from *ndop*.

Another bewitching ritual cloth, used by the Dowayo of northern Cameroon, is the so-called 'firecloth', onto whose base of dark indigo-dyed narrow strip cloth are appliquéd images of lizards and cowries. An additional strip of indigo resist-dyed cowrie designs forms a belt at the top. Only a few of these precious cloths survive, and they are loaned out to be worn around the heads of boys at circumcision and of corpses before interment. The Dowayo refuse to offer interpretations of their perplexing iconography on the grounds that this would remove the essence of the supernatural transformations which take place during circumcision and interment.[49]

Masked figure at an annual Ebira festival wearing special indigo and white patterned cloth, *itokueta*, reserved for use as a shroud. It is used here to represent the continuity between masquerades and the world of the dead. Nigeria, 1960s.

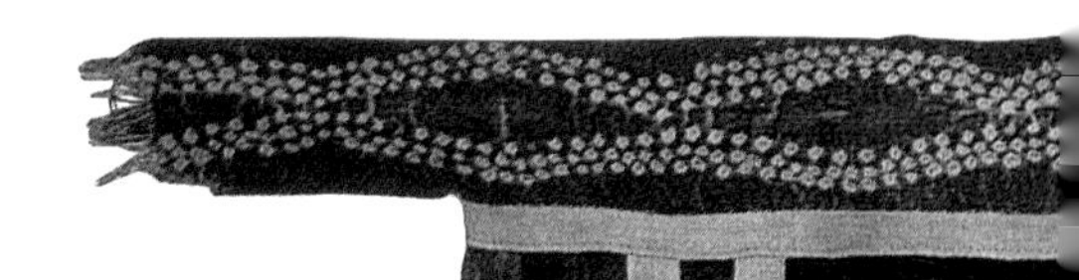

Itokueta, Ebira burial cloth hung on the outer wall of a house whose owner, a person of importance, had died. The cloth would subsequently have been used to wrap the corpse for burial. Nigeria, 1960s.

For the Dogon of Mali indigo-dyed cloth has, once again, strong symbolic associations with rites of passage, although beliefs are changing rapidly as animism is overtaken by Islam.[50] Brett-Smith, who has researched in depth the symbolism of indigo-dyed cloth, suggests that the Dogon classify indigo 'among rotten, fertilizing, and feminine things' and, as elsewhere, the dyed cloth has associations with both life and death.[51] Dogon women wear resist-patterned indigo wraps after marriage, but those worn by the elderly (i.e. non-fertile) are plain. Although the patterns on some wraps are straightforwardly representational (e.g. of a family's onion plots), others are more complex, symbolizing marriage and fertility, ancestral links or aspects of the sacred masked dances that take place at the extraordinary *Sigui* ceremony held every sixty years. If a woman's wrap combines lines of different-coloured embroidery with stitch-resist decoration, this indicates her elevated status as a medicine woman involved in childbirth and female circumcision (see p.221).

At funerals in the animist tradition, a man puts on indigo-dyed trousers while a woman wears an undecorated indigo wrap and shawl over the right shoulder. Most interesting is the shroud reserved for the burial of a village chief, *hogon*. This special cotton blanket, *munyure*, woven locally, bears a

Ceremonial *ndop* burial gowns belonging to Chief Njiké III, displayed by his retainers. The narrow-strip cotton cloth is decorated with an oversewn raffia stitch-resist technique before being dyed in indigo. Bangangte, West Province, Cameroon, 1980.

Rare ritual 'firecloth' used by the Dowayo of Cameroon for ceremonies.

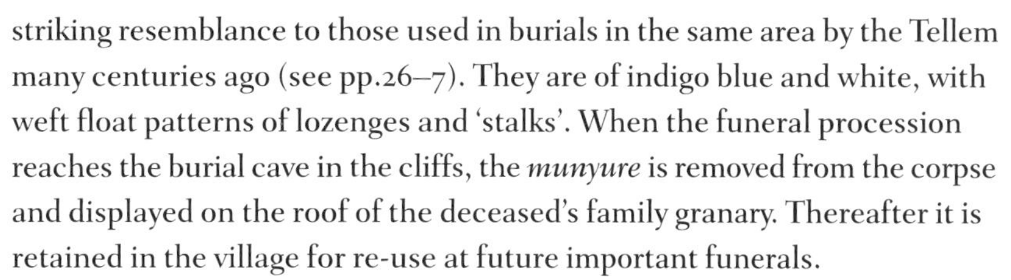

striking resemblance to those used in burials in the same area by the Tellem many centuries ago (see pp.26–7). They are of indigo blue and white, with weft float patterns of lozenges and 'stalks'. When the funeral procession reaches the burial cave in the cliffs, the *munyure* is removed from the corpse and displayed on the roof of the deceased's family granary. Thereafter it is retained in the village for re-use at future important funerals.

In both Senegal and Mali, where the use of indigo is much reduced by the availability of every conceivable colour, indigo owes its survival to ceremonial requirements. Dyers along the Senegal and Niger rivers still produce resist-dyed indigo gowns and shawls for such occasions as a young man's graduation or the obligatory visit a bride pays to her in-laws after her wedding.[52] Even Dakar's famous popular singer Cheikh Lô changes into traditional indigo-dyed robes when travelling to pay homage to his Muslim *marabout*. In regions surrounding Timbuktu indigo blue is used instead of natural black wool in special dramatic handwoven Fulani and Touareg wedding blankets known as *arkilla*.[53]

In Ghana and neighbouring Ivory Coast both red and dark indigo cloths have long been used for mourning (although among the Ashante indigo blue is also related to love, female tenderness, the early dawn and the crescent moon).[54] In both countries modern factory-printed cloth designed for use at funerals is still predominantly blue.[55]

A line of dancers wearing *njop* skirts at a 'cry-die' ceremony staged by the family of a dead man. Babete, near Mbouda, western Cameroon, 1980.

Arkilla kerka, Fulani wedding blanket used to divide the nuptial bed from the reception area in a traditional domed dwelling. At least 5 metres long, made from seven strips about 25 cm wide in wool and cotton patterned with a supplementary-weft technique. The dark colour is deep indigo blue. Niger bend, Mali.

Detail of a woollen ceremonial poncho worn by a Mapuche chief. Southern Chile.

South America

Here, too, coloured textiles played a central role in ritual and ceremony (see pp.19–22). A particularly impressive group of indigo-blue ones are the geometric ponchos and horse blankets of the Mapuche people of southern Chile, which have been described as follows:

The monumental blue-and-white geometric forms of Mapuche chiefs' ponchos must have had a striking visual impact. Radiating power, they proclaimed at a glance the identity and status of their owner . . . Like all Andean textiles, they eloquently and elegantly transcended their utilitarian role, mediating between the living and the dead.[56]

The distinctive textile arts of the Mapuche people, though influenced by the Spanish, incorporated the archetypal Andean design of the stepped pyramid. Plain and skilfully tied ikat yarn dyed in indigo, the colour of infinity, was essential for the production of the most valued Mapuche textiles.

Europe

Among Europeans, 'navy' (indigo) blue clothing is still a common choice for family occasions, particularly funerals. Certain Calvinist centres in the Netherlands were notable not only for the indigo skirts and 'blue-print' scarves generally worn for mourning, but also for the fact that different periods of mourning were indicated by the amount of indigo appearing in clothing and accessories such as belts.[57] Here, as in central Europe, the most elaborate 'blue-print' clothing was reserved for high days and holidays. In Slovakia the enormous trouble taken by the secretive 'blue-printers' when making women's ceremonial shawls is recorded by Vydra. These were printed by hand on both sides using coloured printing paste and exceptionally dark indigo.[58] The embroidered waistbands of *Blaudruck* aprons reserved for religious occasions were also full of symbolism and, like the treasured pleated blue-and-white *Blaudruck* skirts, are still occasionally worn by elderly women in remote areas.[59]

North America

How can one leave the subject of ritual clothing without mentioning blue jeans however everyday they have become? One American writer vividly recollects his teenage years in post-war America, the days of 'junior-high kids sitting on the stoop discussing the coolest way to shrink and fade your new blue jeans', when 'the best thing of all . . . was to bury a pair of jeans in a horsetail patch for a month or so.' The writer reflects:

It never occurred to us that all our sanding and hosing down, bleaching and grating and scalding was a form of adornment in itself, as ritualized as the cutting of tribal scars, as strict and exact as the magic of painting one's body for war or for mourning. It was just what you did with Levis. It was magic all right.[60]

Everyday textiles

In many cultures worldwide indigo dyestuff formed for centuries the bedrock of everyday clothing and furnishings, quite apart from its even more basic use as laundry 'blueing'. Plain blue or a blue-red or blue-white combination were standard colours in many rural societies, since blues and reds were the world's basic colourfast natural dye colours. Indigo's affinity for all fibres gave it an advantage. Madder and the insect reds dyed silk and wool easily, but cotton fibres needed lengthy processing to accept these dyes.

Historically, the blue-red combination held pride of place in, for instance, many rural areas of Central America and in Indonesia. Soft madder reds and indigo blues of all shades characterize tribal rugs of central Asia, while crimson reds (made from insect dyes) and midnight blues were favoured by the Islamic courts.

As for combinations of blue and white, these predominated in West Africa, Japan, China and her neighbours. The enthusiasm aroused far and wide by China's 'export blue-and-white' porcelain did a lot for indigo, and maybe blue-and-white textiles helped to maintain that popularity. In China itself rugs produced during the Ming dynasty (1368–1643) consciously echoed the blue designs of Ming porcelain.[61] In the West the arrival of mass-produced Chinese blue-and-white porcelain in the following Qing dynasty,[62] coincided with the appearance of indigo, cotton and resist techniques. People of all social classes in Europe could now enjoy the fresh and flowery look of blue-and-white printed cotton. The bourgeoisie could wear their *toiles de Jouy* chintzes, while the humble could afford *Blaudruck* 'blue-prints'.

Japanese woodblock print (two halves) of a group of people in various indigo- patterned robes. The man on the far right wears a double ikat kimono, the woman at his feet and the girl wear stencil-patterned cloth. Sashes characteristic of late eighteenth-century dress.

Detail of a late nineteenth-century *Shirvan saph* with seven *mihrabs*, prayer niches. Four are red, two are dark indigo, the other is bright indigo blue.

Even with everyday *Blaudruck* there was more to patterning than meets the eye. In central Europe it could be a directly religious statement – flamboyant designs of white-on-blue for Catholics, simple white or even blue-on-blue for Protestants. With the latter, blue-on-blue patterning was sometimes so discreet that it could hardly be detected, but it was essential to wear *Blaudruck* rather than plain cloth as it had more status.[63]

Absolutely plain and simple indigo-dyed cloth made ideal workwear everywhere. When it grew tired, it could be freshened up by returning it to the dyers for a new dip. In much of Asia belief in indigo's medicinal and protective properties added to its popularity. Nineteenth-century travellers to the East took for granted the sight of peasants toiling in the fields in their clothes of blue; and the blue-robed Egyptian *fallahin*, as depicted by Orientalist painters, epitomized the timelessness of subsistence living along the banks of the Nile.

Rural men in indigo shirts and trousers in a mural painting at Wat Chai Sri, north-east Thailand.

Likewise workers in Europe also wore blue for centuries. French farmers and butchers wore dark linen smocks, a British gardener was known as a 'blue apron man' and trousers in Holland were blue (hence the expression 'enough blue sky to make a Dutchman a pair of trousers'). Germany's 'Blue Monday' is thought to relate to the habits of the woad-blue artisan.[64] The list is endless. In urban centres, too, though the more affluent had a wide choice of colours, blue was often popular, whether for a kimono in Japan, the robe of an Eastern aristocrat or the dress coat of a European.

Europe

Although the popularity of red as a prestige colour gained ground in the high medieval period, it is hard to imagine how the cloth manufacturers would have managed without woad, both as a single colour and as a statutory base for many more.[65] Even when there was a wider choice of dyes available, draperies and tapestries in houses of the better-off were awash with soft blues or greeny-blues. But people who could afford brighter colours enjoyed wearing them. In England different colours seem to have been worn at different seasons. Late fourteenth-century records at York show blue predominating between the months of September and December, but giving way to russets in the spring, although blues, 'plunkets' (common pale blue clothes) and greens were still being worn.[66]

In Britain, as in the rest of Europe, blue garments were a distinguishing feature of many groups of people. The Ludlow Palmers, a religious group from the town of that name, can be seen wearing their traditional blue coats in a Ludlow stained-glass window that depicts their pilgrimage to Jerusalem. Shakespeare refers to a beadle as a 'blue-bottle rogue',[67] because in his day beadles, constables, policemen, almsmen and others were all required to wear blue. Scholars at charity schools were known as 'blue-coat boys', and the monarch used to distribute blue cloth to the poor on Maundy Thursday. 'Bluestocking' refers to serious-minded women who chose to wear woad-dyed worsted stockings rather than fine black silken ones.

Nothing could be more hard-wearing for police and service uniforms than durable indigo-dyed serge. Even in the early decades of the twentieth century such British cloth was advertised as being dyed with 'Guaranteed indigo fast dye resistant to sun, sea and air'.[68] When it was intended for use in the colonies it was subjected to rigorous inspection by a Crown Agent inspector. Even after the Second World War railway and post office workers were dressed in indigo-dyed woollen cloth.[69]

Rear-Admiral Sir George Cockburn wearing an indigo-dyed naval uniform of a style worn between 1812 and 1825. Portrait by John James Halls, 1817.

Back in the Middle Ages Genoese and Dutch sailors wore trousers made of hard-wearing fustian, a coarse twill, usually dyed blue, and by the first half of the eighteenth century indigo blue was a standard colour for many European naval uniforms. It only became so in Britain in 1745 when some naval officers petitioned the Admiralty on the subject. They each suggested different colour schemes, but an indigo 'navy blue' uniform with white facings was chosen in order, so the story goes, to please King George I, as it was a colour combination worn out riding by one of his favourite women, who also happened to be the wife of the First Lord of the Admiralty![70]

Armies, too, kept all those in the indigo industry busy. From the eighteenth century on much of the indigo produced in the West Indies and America was used to dye uniforms for the large armies of Europe, not least Napoleon's. In the early days of Germany's synthetic indigo industry her own army was a major consumer,[71] and Britain's requirement for indigo to dye service uniforms (and 'hospital blue') lay, as recounted on p.83, behind her desperate need to manufacture her own synthetic indigo during the First World War.

Asia and the Near East

The diversity and beauty of Asian textiles, even those worn daily, have entranced travellers, ethnographers, textile historians and collectors for centuries. One of the most fascinating aspects of those of Southeast Asia is the way outside influences, coming from India, China, Islamic countries and Europe, have been successfully blended with existing indigenous models. The visual messages contained in cloth colour and design, though almost forgotten in most societies, can still be unravelled in more isolated communities. In her study of a remote village in eastern Java in the late 1980s Heringa identified the metaphorical meaning of the changing coloured cloth worn in a person's life. As one simple example, it is only when a young girl becomes a mother that she is allowed to wear a daily shoulder-cloth of 'black', for the mingling of red and blue dyes used to make it are considered symbolic of the physical union of a husband and wife.[72] Similarly, in one part of Flores Hamilton describes the progression of clothing worn until quite recently by a man during his life. As a child he would wear a plain sarong until he was circumcised. Then, after a three-month period of isolation in the forest, he would return to the village wearing a sarong and shoulder-cloth made from indigo-dyed material patterned with simple ikat motifs. These clothes signalled his marriageability and later became part of his daily dress.[73]

In India any traveller will immediately notice that everyday clothing is a riot of colours, the most dominant being not blue but red. This may seem strange in a country so famed for the export of indigo. Much indigo was indeed consumed within the country, but it tended to feature in patterned textiles (printed *ajrakhs* and painted *kalamkaris*) and in cloths containing mixed colours, such as green, whose dyeing required indigo.

In Japan, on the other hand, despite the ability of her dyers to produce an extensive range of colours, indigo-dyed clothing and furnishings, both plain and patterned, were standard in rural areas from the beginning of the eighteenth century. Almost all the country people pictured in Hiroshige and Hokusai woodblock prints are wearing indigo. However, urban dwellers, too, had a taste for indigo, and in the late Edo period many of the same styles and patterns were popular in both the town and country and most of these textiles were dyed a deep blue.[74] Amongst the various domestic furnishings so dyed the futon cover was particularly prized. It was usually made of either indigo-dyed *kasuri* (ikat) cloth or fabric on whose indigo ground were stencilled designs symbolizing happiness in marriage.[75]

In China country clothing must have been just as blue long before its political enforcement by Chairman Mao in 1929, but natural indigo had given way to the German synthetic substitute in the early years of the twentieth century. Today plain, embroidered, and resist-patterned indigo-dyed clothing survives only among the ethnic minority peoples along China's southern borders.[76] The wearing of shiny indigo turbans by men in certain villages is reminiscent of the very different cultures of southern Arabia and West Africa.

As for China's various types of carpet, one study emphatically declares: 'The richness and various shades of the blues on Chinese rugs are undisputed and unparalleled among handmade rugs.'[77] Moving westwards, rich indigo blue features as a predominant colour in many Tibetan rugs, including a rare seventeenth-century example.[78] Tibetan dyers may have found indigo hard to use, for their rugs always have an *abrash*, mottled look, that indicates the difficulty encountered with dyeing their lanolin-rich wool.

Still further west in central Asia dark blues were preferred to light as they were a sign of wealth. An interesting design emerged in nineteenth-century Persia based on geometrical forms woven in the contrasting tones of such dark indigo blue and ivory white that they have been called 'black and white'.[79] Textiles in this colour scheme were made in many parts of the country, not least the trappings that were originally woven for nomadic use by the Qashqa'i in the Fars Province.

Clothing in central Asia, the Near East and North Africa relied heavily on indigo by medieval times.[80] Among urban societies changing sumptuary regulations issued by the Muslim authorities would stipulate specific colours for different confessional groups (such as restricting Christians at one period to wearing a blue turban, or at another requiring Jewish converts to Islam to wear blue garments with wide sleeves).[81] Women's indigo-dyed outer wraps remained a common sight in the cities in the first half of the twentieth century, whatever colours were worn underneath.

Padded Japanese fireman's helmet of indigo-dyed cotton with stencil-resist patterned lining, worn with a thick quilted jacket. The flap was pulled down before a blaze was tackled and the whole outfit drenched with water. Late nineteenth or early twentieth century.

Page from a book of paintings illustrating Miao tribal life. The author of the original text was Wang Chin, nineteenth century

Among rural Arab communities, the best known female garments made of indigo-dyed cloth are the beautifully embroidered and appliquéd coats and dresses of Palestine and Syria (see pp.171 and 173). Women throughout southern Arabia too wore (and some still do) dresses of polished indigo cloth, sometimes decorated with applied brass 'sequins', amulets, cowries and beads, in addition to embroidery. The Danish explorer Carsten Niebuhr, passing through the Yemeni Tihama in the eighteenth century, observed that the veiled women 'all looked as if they had been dipped fully clothed into the great earthenware vessels on the outskirts of the town where indigo is made'.[82] Indigo-dyed face masks of Arabia and the Gulf came in all shapes and sizes, and the indigo rub-off was welcomed as it was thought to 'whiten' the face. Intended to conceal the features, the shaping of these masks could in fact increase their allure.

As for the men, one traveller to Yemen at the beginning of the twentieth century stated quite baldly that 'indigo blue is the national colour'.[83] Until the 1970s the tribesmen of southern Arabia were proud of their indigo-stained skin (see pp.200 and 224), and their wraps and head-dresses were saturated with surplus dye to reinforce the stain. Nowadays indigo-dyed garments are rarely worn on a daily basis, but may be put on for state occasions. Some men, however, and not just the elderly, still sport the traditional shiny turban, as often as not with a sprig of aromatic herbs stuck into the top (see p.134).

Miao man in a traditional indigo-dyed long coat with decorated purse, his hair bound in a top knot. Biashia village, Guizhou, south-west China, 1993.

Africa

The Arabian habit of wearing indigo turbans clearly spread to West Africa with Islam. At the Fulani courts of northern Nigeria particularly fine ones were worn for Muslim celebrations,[84] and equally voluminous ones still enjoy the highest reputation among the Touareg of the Sahara where they are known as *tagelmoust*. Other 'blue men', notably camel-trading Moorish groups of southern Morocco and Mauritania can still be seen wearing indigo-dyed turbans and robes, especially at festivals.

Islam is also thought to have introduced embroidery, often practised on a ground of dark indigo-dyed cotton or on handwoven strip cloth in shades of blue. Garments of all sorts were embroidered, usually by men, the most outstanding being wide-sleeved gowns. Many examples from Nigeria and neighbouring countries are in the British Museum collection.[85]

In the sixteenth century the traveller Leo Africanus reported that notables in the northern Sahara were distinguished by the wide-sleeved robes traded from further south.[86] Barbot in the seventeenth century gave the impression that indigo-dyeing was a common activity even then,[87] but production clearly expanded until more and more people could afford to

A Touareg man encountered at El Mecki in the Aïr mountains of Niger, 1970.

wear indigo-dyed clothing. Yet, in 1865 in middle Niger a dark indigo robe could, according to one historian's calculations, cost up to 10,000 cowries, about ninety times the daily subsistence for a man and his horse.[88] Blue became the predominant colour for clothing of all social classes, but what primarily distinguished the more prestigious garments was the amount of indigo dye used. A tribal chief, for example, would wear a capacious narrow-strip robe made from cloth that had been immersed in the dye vat up to forty times. Those who could not afford indigo-dyed clothing could make white garments resemble the real thing by beating indigo paste, *shuni*, into them or by rubbing indigo on the body to impart a bluish tinge to a gown.[89]

Resist-patterned clothing worn by men and women was tailor-made for draping on the body. Voluminous robes known as *bubus* are said to have been introduced by missionaries for modesty's sake in the nineteenth century, but their outsize necklines slip off the shoulder in a rather suggestive manner. The resist-dyed indigo wraps of the Yoruba, which were widely collected by Europeans, reveal Islamic influence in their designs as well as depicting local flora and fauna and Yoruba mythology. During the colonial period the repertoire of traditional motifs expanded to commemorate such topical events as the Silver Jubilee of King George V in 1935.[90] Until the 1960s a large Yoruba cloth market like that of Oje was a sea of blue *adire* and woven cloth. Thereafter a wide choice of colourful clothing became available, and stalls in every market of West Africa are piled high with Western *fripperie*.[91] Yet even today the Yoruba dyer Nike Olaniyi-Davis has no doubt about the place of indigo in the hearts of women in her homeland: 'Real indigo is the colour of love – cloth dyed with it has its own *sound* and smell.'[92]

The East African mainland south of Sudan does not appear to have had an indigo-dyeing tradition of its own, but in the past Muslim traders exported blue-dyed cloth from India and Arabia expressly to be unravelled and incorporated into locally woven cloth.[93]

Yoruba *adire* resist-dyed cloth for sale in a street market of Ibadan in the 1960s.

E 5/588

Central and South America

Across the Atlantic, despite the preponderance of red in woven and embroidered textiles of wool and silk, indigo was the most widely used dye in traditional Indian cotton skirts and shawls in Guatemala, El Salvador, Ecuador, Mexico, Peru and Chile. These could be either of prized ikat weave, *jaspeado*, or cheaper plain dyed cloth with embroidered stripes.[94] In the nineteenth century the beginning of the indigo-making season was even the occasion for a large festival which included as entertainment a 'bull and horse' dance.[95]

In Mexico some traditional communities cling on to the use of indigo, even though other natural dyestuffs have been abandoned.[96] The weavers of Chiapas, for example, still like to incorporate indigo-dyed stripes into woven skirts and to dye skirts in indigo for the Feast of All Souls; and the ubiquitous blanket, *sarape*, of Oaxaca still features much indigo.

Tzotzil girls wearing indigo-dyed wrap-around cotton skirts with red sashes and everyday tops, *huipiles*, embroidered in red. San Andrés Larrainzar, Chiapas, Mexico, 1978.

One fascinating Mexican survival is the ancient custom of weaving highly valued skirts whose warp stripes are made from cotton dyed with shellfish purple and indigo, and silk dyed with cochineal (now synthetic).[97] Their high price reflects the laborious procedure required to 'milk' the glands of the molluscs *in situ* directly onto cotton yarn without harming them. The molluscs are quickly returned to their rocks on the Pacific coast where they can be 'harvested' again a month later. This is the only place in the world where purple dye has been extracted from shellfish without harming them. Though recently badly depleted by greedy foreign entrepreneurs, the molluscs are now under government protection and projects are even under way to use them as a source of natural paint pigment.

Blue denim – a global phenomenon

'A well-fitting pair of denim jeans is the most important – and often the most sexy – thing in a girl's wardrobe.' Thus declared my student daughter in 1998 on purchasing a skin-tight pair of blue jeans, demonstrating that, fickle as fashion can be, one indispensable item connects her youth with her mother's in the so-called 'swinging sixties'.

What are the origins of the most extraordinary fashion story of all time? Although its roots are obscure, there are two semantic clues. One leads to the Italian port of Genoa, the other to Nîmes in southern France.[98] In early Islamic times a textile known as fustian, whose name suggests it came from the city of Fustat (old Cairo), was exported from Egypt to Italy. In the Middle Ages the Genoese manufactured their own version, a tough cotton or linen cloth known elsewhere as *Gene fustian*. When dyed, it was with indigo blue the colour that typified practical workclothes, including sailors' trousers. Other European weaving centres followed suit, and it is thought that the name *Gene fustian* was abbreviated in English to *Gene*, hence 'jeans'.

As for the second clue, Nîmes, a major centre of cloth manufacture, produced from the end of the Middle Ages its own cheap everyday cloth, a twill called *serge*. This *serge de Nîmes* was imitated in Britain under the name of 'Nims'. In Nîmes the original woad-dyed serge was mainly of wool, but later hemp, silk waste and cotton were included. In the eighteenth century imported West Indian indigo enabled much cotton to be dyed in the new woad/indigo vats of Nîmes and traded overseas (often by Genoese merchants). During recent archaeological excavations beside the Nîmes canal, a group of wooden vats, dated to around 1700, were discovered in the old dyeing quarter that still had traces of blue dye in them.[99] The ancestors of modern jeans could have been dyed in these very vats. The names of these two types of practical cloth, *serge de Nîmes* and *gene fustian*, found their way into a classic American tale of rags to riches.

This story begins in 1847 when Levi Strauss, an eighteen-year-old Bavarian, emigrated to New York.[100] Six years later, Levi moved to San Francisco to open a branch of the family firm selling wholesale dry goods. Bolts of cloth were sold to an immigrant Latvian tailor, Jacob Davis, who made practical 'waist overalls' (trousers). Finding that the pockets of these garments tended to tear at the corners, Davis hit upon the idea of strengthening them with copper rivets. However, he could not afford to take out a patent, so he invited Levi Strauss to become his business partner. The pair opened a factory in 1873 and produced the new waist overalls using indigo-dyed twill from an American mill. These forerunners of today's jeans were a great success and other trademark, now familiar worldwide, were later registered. They included the double line of orange stitching on the back pocket, known as 'the Arcuate', which has become the longest running trademark in the United States.

Having brought to light a product that has hardly altered in a hundred years, Levi Strauss died in 1902. After his death the market for denim trousers, now made from cloth manufactured in California, grew and grew. In the 1930s the 'fashion' element of blue jeans was spotted by Vogue magazine, who introduced them to America's East Coast. Soon afterwards the Second World War began, and jeans reverted to their purely functional role, as they were declared 'essential commodities' by the US government. During the war American GIs gave Europeans their first sight of blue jeans. 'The rest is history.'

In 1960, when Levi sales topped 46 million US dollars, the word 'overalls' was replace by 'jeans' in advertising. Jeans became all the rage and inevitably Levi Strauss has attracted many competitors such as Wrangler and Lee. However, since 1970 the company has expanded into many countries, opening 'Levi's Only Stores' in Europe and the Far East, and global sales are now worth several billion US dollars.

Wooden woad dye vats excavated beside the canal in the old dyers' quarter of Nîmes. Late seventeenth or early eighteenth century.

And why did jeans take the world by storm? Glamorized by the GIs, second-hand GI trousers after the war became desirable cult objects in Europe. Demoralized by years of conflict, the youth of Europe aspired to emulate all things American. Blue jeans became their icon, particularly after their adoption by film stars and pop musicians including Marlon Brando, Marilyn Monroe, Elvis Presley and the Beatles. Denim jeans appealed to the young as they embodied not only the American dream but a rebellion against their parents and the establishment. When the first Dior model dared to wear them, it was thought to give her a shocking whiff of 'St Germain' (these were the days before St Germain became respectable).

Of course the young grow older! And maturing with them are their jeans, now firmly straddling the work ethic and mainstream fashion. They have acquired an international and egalitarian status appropriate enough in a world being unified by modern technology. Blue denim is now promoted by fashion pundits as an ultra-contemporary product with historic roots going back to the Wild West, so that one typical advertisement declares: 'You can be a computer programmer in Copenhagen, but when you pull on your jeans, you hear a horse whinny and feel a guitar on your back.'[101] Even patched and slashed jeans, which originally emphasized a rebellious attitude, have been hijacked by high fashion. Buddhist adherents in China and Japan, like the followers of the Mahdi in Sudan (see pp.80–81), wore patched clothes as a sign of humility, but soon the quality of the patches themselves conferred rank.[102] Similarly, today's fashion designers often

Counter card of the 1950s advertising Levi's waist overalls.

have a field day with patched and faded denim, originally treasured by the young and considered to improve with age.

It is these unique fading qualities of blue denim that have preserved indigo for the mass market. Without the craze for jeans other synthetic blues would have long since taken over. The twill weave of denim is prone to overall abrasion on its most exposed parts, and this is accentuated both by the fact that only the warp threads are dyed, and by indigo's tendency to rub off. This worn look is so popular that it is deliberately created on new jeans by abrasion (with pumice stone) or enzyme washing in the factory.

Today's blue denim is produced in a wide variety of fibres and textures, which include combinations of linen and modern fabrics such as viscose, tencel, lycra and rubber-backed indigo cotton for rainwear. Indigo-dyed yarn is made into all kinds of knitwear, using different weaving techniques such as jacquard.[103] Yet most are still marketed as 'authentic denim'. The possibilities are endless, and the story of blue denim will surely run and run.

་འདི་སྐད་ཅེས་གསོལ་ཏོ། །བཅོམ

་ཏུ་ཕྱིན་པ་ཡི་གེར་འབྲུར་བྲིས་ནས

་པར་བྱོར། །རབ་འབྱོར་གཞན་ཡང

ས་དང་། ཡུལ་འཁོར་དང་། རྒྱལ་པོ

་པ་དག་འབྱུང་བར་འགྱུར་། ཆོས་

ནང་པ་དག་འབྱུང་བར་འགྱུར་། བྱང

For monkish garb and for the robes of saints . . . first take black with a little indigo, add a little lead to lighten it sufficiently and then shade on it with indigo.
(STRASBOURG MS, FIFTEENTH CENTURY)

Blue Art

For all its pre-eminence as a dyestuff, indigo was also much used from antiquity well into the twentieth century to make paint and ink and to colour leather and paper.

Paint and ink

To dye cloth, indigo needs to be chemically converted by the vat process, but the extracted dried pigment in its oxidized state, sometimes referred to as 'indigotin', can also be ground up finely like a mineral and mixed with a medium to make ink or paint. Different concentrations of pigment and various preparation methods can produce colours ranging from blues as bright as lapis lazuli to soft bluey greens, although the more the pigment is extended the less light-fast it becomes. Some base ingredients, such as egg white, enhance luminosity. For paint, as for textiles, nature does not provide green colorant. The green of man-made verdigris tended to blacken and was corrosive, a fact of which medieval artists and craftsmen were clearly aware. A common combination therefore, as with dyeing, was to mix blue indigo pigment either with a vegetable yellow or with orpiment (derived from arsenic).[1] Although it was also combined with oil mediums, indigo's main use was for fresco and water-based paints and inks.

The ancient Mayas of Central America, for whom blue was the colour of human sacrifice, used their celebrated 'Maya Blue' paint (see p.22) to

Detail from a large freestanding page of an early eighteenth-century Tibetan Mahayan *sutra*. The indigo paper is glazed in the centre to create a striking background to the gold Buddhist text.

decorate pottery and temple frescoes. It is now known that it was made by combining indigo with a white clay mineral.[2] The Spanish much admired the quality of paint pigments used by the American Indians.[3] Recent analyses have revealed other applications of indigo paint in that part of the world. For example, it was used, as was shellfish purple, to highlight details in outstanding cloth paintings of pre-Hispanic Peru, whose exuberant mystical designs were otherwise of restrained earth colours.[4]

In the Classical world, according to both Pliny and Vitruvius, expensive indigo pigment, imported from India, was used by painters of frescoes, who admired its purplish hue, but had to beware of counterfeit versions.[5] A learned discourse on the nature of indigo pigment in Classical times appeared in Beckman's nineteenth-century *History of Inventions*. Although Egypt has often been described as the founding country of Western chemistry and alchemy (the word has an Arabic derivation), during the Classical period in the West the chemistry of metals, dyes and medicines was clearly developing in many parts of Asia too.[6]

By early medieval times scribes and manuscript painters of the East had a broad understanding of the chemistry of paint ingredients, and probably selected indigo for its non-corrosive character, using it in the preparation of various coloured paints and inks.[7] Indian and Persian manuscripts often have wide borders of plain indigo blue in addition to blue details,[8] and indigo is commonly found in various mixtures in Islamic manuscript illuminations elsewhere.[9] It was combined with orpiment for various blue-greens, with soot for darker shades and with lead for sky blues. The treatise of Ibn Badis, written in the first half of the eleventh century AD, is a major source of documentation on the subject of Arabic bookmaking.[10] Several of his recipes for preparing subtle hues of inks and *liqs*, which were ink-soaked wads for use with the pen, include indigo; he mentions combinations with such substances as red lead, purified celandine, gallnuts, saffron, coriander and arsenic.[11] Further east, in China, indigo was the standard pigment for blue ink. By the tenth century AD some inks were commonly used in Chinese medicine, thus demonstrating the difficulty of pigeonholing the functions of natural dyestuffs.[12] Much later, woad was even used to darken green tea in northern China to suit the export market to Europe and America, hence the nickname 'tea indigo' given to woad by Western traders.[13]

In medieval Europe oriental indigo was still being imported mainly as a paint pigment before it ousted woad as a dyestuff, although its exact nature and its relationship to woad still caused confusion.[14] (Even today, when historical samples are analysed by scientists, woad and indigo pigment cannot be told apart.) Painters needing a cheaper source of indigo blue made use of the scum, or 'flower' – known as 'florey' (*fior de guado*) – that floated on the top of a fresh woad (or, later, indigo) dye vat, or even used recycled woad dye recovered from woollen cloth shearings.[15] After the Middle Ages woad paints were sometimes referred to as indigo,[16] and various sources provide recipes for their manufacture until the seventeenth century, when woad became obsolete as an artist's colour.[17] Egg white, urine and vinegar were among the recommended binders to mix with woad/indigo, while 'azzurro', a sky blue colour, could be produced by using such ingredients as calcined marble powder, crushed egg shells and wine. Woad was also made into soft crayons for drawing, hence the term 'pastel' (see the Glossary) now used for all such coloured crayons. Woad/indigo paint was certainly used in such medieval decorative features as cloth-hangings.[18]

Before its wide-scale importation for the dyeing industry from the seventeenth century, imported indigo pigment (usually called 'Baghdad' in painters' manuals)[19] was expensive, and often taken to be a mineral. It was, though, not as luxurious as the other main medieval blues that were indeed obtained from exotic minerals – the highly esteemed lapis lazuli (ultramarine) and azurite – although it could be used to imitate them. Cennino Cennini in his late fourteenth-century manual advises: 'If you want to make a drapery in fresco which will look like ultramarine blue, take indigo and lime white, and step your colours up together; and then, in secco, touch it in with ultramarine blue in the accents.'[20] The medieval artist used indigo either for shadows and relief or for a base colour.[21] To brighten the blue, indigo was generally mixed with lead white or lime white, extended with various substances as described above.[22] It could be applied to all supports, including paper, wood, cloth or walls. Cennino Cennini specifies many other recipes with indigo, including some for popular green mixes with orpiment, and purples with haematite (iron oxide ore).[23] Other combinations used by painters include indigo and brazilwood for purple, and indigo with red lead and carbon to enhance blacks for manuscript lettering and illumination.[24] As more paintings and manuscripts are analysed, particularly using the latest non-destructive *in situ* methods, it will be easier to calculate the relative popularity and different uses of the various blues available at the time.[25]

From the seventeenth century the prices of ultramarine and azurite blues were frequently so exorbitant that indigo, now much cheaper, was often used instead. Although not ideal as an oil colour, owing to its tendency to fade despite elaborate preventative treatment, indigo was nevertheless commonly used by artists such as Rubens and Frans Hals to eke out the more expensive blues.[26] Such painterly secrets are being revealed by microscopic analysis. In the early eighteenth century Prussian Blue was invented by accident; it soon replaced indigo for oil painting, and later functioned as a textile dye.

Among watercolourists and ink manufacturers, however, indigo, which flowed well, continued to be widely used. A chart of permanent colours, produced for professional painters in 1816, lists three blues, Prussian, the new cobalt and indigo. Indigo, it says, makes subtle shades:

Twilight and Evening Skies but not sufficiently bright for skies on clear days; is useful for the green of trees when mixed with Burnt Sienna and Gamboge; useful when mixed with Lake and Gamboge, to make Greys and Neutral Tints with; washes smoothly much easier than Prussian Blue, and probably on that account often used, where Prussian Blue would be most proper.

The chart suggests indigo in other mixes, too, for different greys, greens and purples. Indigo-based watercolours survived into the twentieth century – Payne's Grey, so beloved by landscape artists, was made from a mixture of lake, raw sienna and indigo.[27] As for ink manufacture, a late nineteenth-century German source declares:

Until the artificial dyes were invented indigo and cochineal carmines had the greatest colouring power known, and both were therefore much used by the ink maker in spite of their high price. In fact, no other dye is so suitable for this purpose as indigo carmine, but we now have artificial dyes . . . and they have therefore replaced it to a large extent. It is nevertheless suitable for every kind of ink, and in addition forms in itself an excellent ink both for writing and for stamping. We must hence describe it carefully.

The author provides recipes for the preparation of indigo carmine with sulphuric acid, and many inks based on this (as well as recommending it for laundry blue).[28] Although rated 'permanent', indigo paint was not altogether colourfast, especially when extended with white.[29]

(m.2) Week 12: the time changes In wages. In 2 carters 2s. In one labourer in the quarry 11d. In 5 other labourers there 4s 2d. One lb of azure (*azur'*) bought in London by the lord [bishop?] (*empt' London' per dominum*) 3s 6d. One lb of indigo of Baghdad (*inde baudas*) 18d. 4 lbs of verdigris ('*verdegris*') 2s 4d. 4 lbs of vermilion ('*vermilioun*') 2s 8d. 5 lbs of white varnish (*werniz alb'*) 5s 3/4d. 3 quarters (*quarteroniis*) of 'cinople' 4s 9d. For 1000 gold [foils] 38s 4d. 6 lbs of white lead ('blamppl') 18d.

Total £4 14s 5 3/4d.

An entry from the Exeter cathedral fabric accounts of 1320–21 which includes 'Baghdad indigo' in the list of paints bought for the altar decoration.

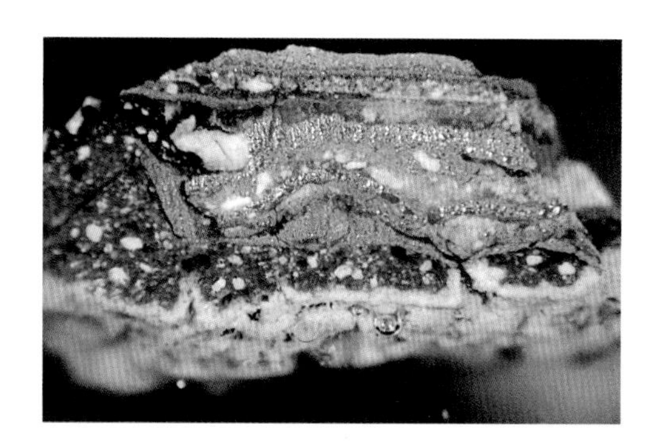

[above]
Microscopic picture of a small fragment (0.75 mm long) surviving on a moulding on the tympanum of the central porch at Salisbury cathedral's west front, showing original indigo scheme and later repaints.

[above left]
The ceiling of the reconstructed Shakespeare Globe theatre during decoration in 1997. The indigo blue of the painted heaven is offset by gilded mouldings (gold leaf is being applied on the right of the picture).

For architectural polychromy indigo functioned in much the same way as it did for artists. Again, analysis of paint samples is backing up literary references. For example, recent examinations of tiny fragments of paint from British cathedrals, which have survived centuries of neglect after the Reformation as well as the zealous cleaning of the stonework early this century, have confirmed the use of relatively expensive indigo.[30] It has been found at both Salisbury and Exeter, and also in contemporary Norwegian altar frontals.[31] It features in numerous accounts of expenditure for decorative schemes, including Exeter cathedral's medieval fabric rolls, where it is called *hindbaud* and *inde baudas* ('Baghdad indigo').[32] Indigo was most useful for emphasizing relief in wooden and stone carving.

Despite not being ideal in an oil medium, indigo was nevertheless the best, affordable oil-based blue for house painting before the appearance of Prussian Blue in the eighteenth century, and it continued to be mixed with distemper to create house paints of subtle colours, such as 'Chinese Green', well into the nineteenth century.[33] (In Tingry's authoritative house-painters' guide of 1830 there is a surprising recipe for using woad to make a yellow paint known as 'pink', the noun that then described a yellow 'lake', i.e. a pigment made from a dye and a mordant.)[34] In German Thuringia woad paint, said to be waterproof and a wood preservative, was used to decorate woodwork in churches and houses, and in the 1980s Wolfgang Feige of Neudietendorf, near Erfurt, developed a solvent-free version. In 1997 indigo paint resurfaced unexpectedly in London when natural Indian pigment, ground up coarsely to give a velvety look, was used by the bucketful for painting the heavens over the stage of the reconstructed Shakespeare Globe theatre.

Some people, including the Yoruba of Nigeria, added left-over indigo dye to house plaster, often applied around doors and windows.[35] The blue tint was said to deter insects. In the East indigo pigment has been used in a variety of decorative ways. In Burma it appears as a colorant for wall-paintings and lacquerware.[36] In India it was often used to paint temple hangings, *pecchava*, partly because the colour itself symbolized the god Krishna,[37] although when indigo is used for painting onto fabric rather than dyeing it, it does tend to be patchy. The preparation of indigo paint for cloth painting is tricky, but a few Indian artists today continue the tradition, using natural indigo pigment from southern India.[38]

There are hopeful signs of a revival of the use of indigo pigment for paint and ink. There are modern artists in both East and West who use indigo in their work for its symbolic value. Commercially, the food and drug industry began in 1998 to pioneer 'environmentally friendly' printing ink on packaging materials, using indigo pigment extracted in new ways from English woad (see pp.100–102 and 113).

Outer cover of a luxury Korean illuminated manuscript of a Buddhist *sutra* of 1341. Painted in gold and silver on indigo paper. Such skilled work was highly regarded.

Colouring leather, parchment and paper

The practice of staining parchment and paper with costly dyes, both for bookmaking and for more general use, also has a long history and seems to have combined practical and aesthetic considerations.

Coloured leather was used by the Ancient Egyptians for hangings and canopies; the background of a decorated leather funeral canopy dating to the 21st Dynasty was stained in blue from an unknown source.[39] The Ancient Greeks stained belts and straps with costly shellfish purple. In the Byzantine Empire pages and bindings of luxury ecclesiastical codices, and later also important secular state documents, were stained purple and inscribed in gold or silver. Islamic calligraphers and bibliophiles also loved coloured manuscripts and bindings. Blue was an appropriate colour for religious texts as it represented the infinite.

Detailed recipes for dyeing, sizing and burnishing leather in medieval Europe and the Near East can be found in various sources.[40] Rosetti's *Plictho* of 1548 (see p.140) covers the tanning and colouring of leather in 'Damascus, Syria, Skopia, Turkey, Italy and Venice'. Typical recipes for making an azure colour instruct the tanners to dress the leather with alum and eggs before colouring it with indigo, normally ground up with whiting and blended with honey or vinegar, wine or lye, and gum arabic. Woad should be mixed with water taken from drains, 'that is, mixed with human urine', and chicken droppings. There is also a recipe for making a green by adding indigo to crushed and dried 'apples of buckthorn'. To bring out the full colour the recipe advises: 'Give them the strop and the button and they will become pretty and lustrous.'[41] In the fifteenth century one of Cennino Cennini's recipes for colouring both parchment and paper blue recommends grinding 'two beans' of Baghdad indigo with half an ounce of white lead, then adding this to a tempera medium.[42]

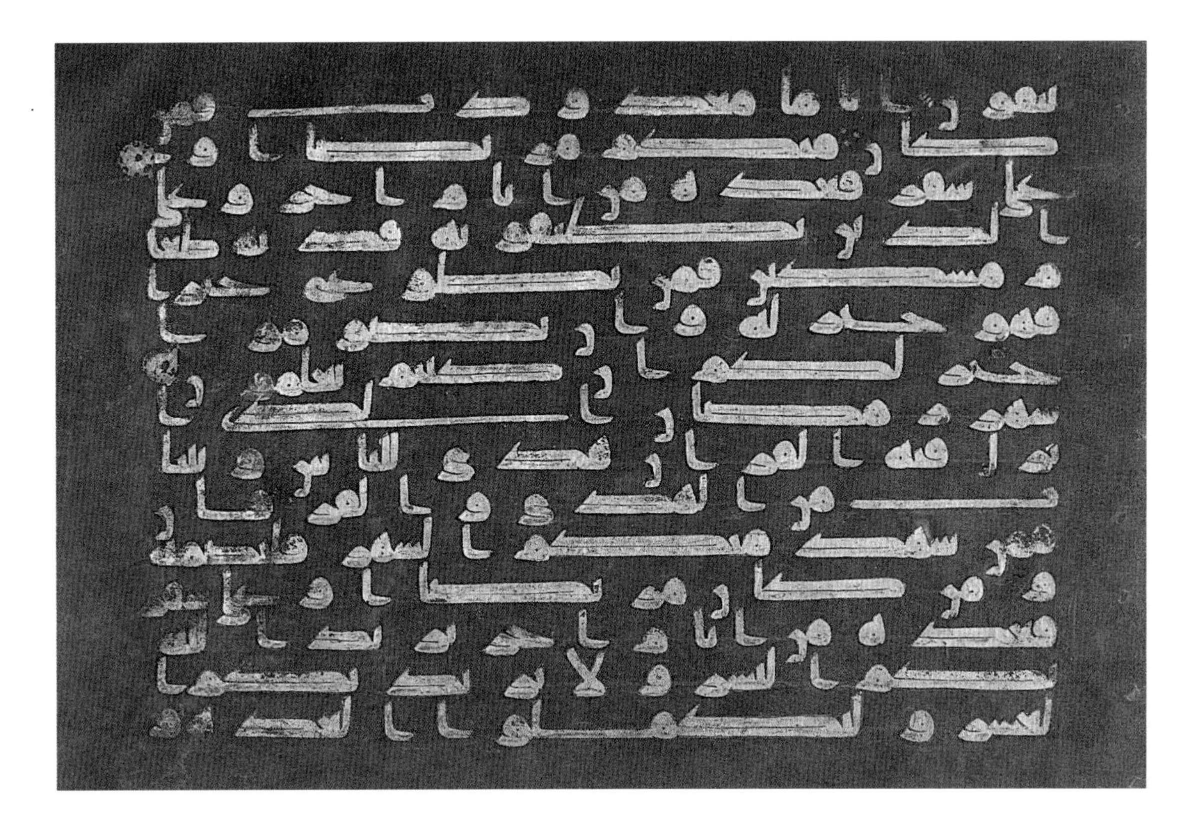

Indigo parchment Qur'an leaf from the Great Mosque, Qairouan, Tunisia, tenth century (28 x 35 cm).

As regards paper, the original reasons for using dyes or stains appear to have been purely practical, to strengthen paper and to repel insects, similar to beliefs in the durability and protective qualities of cloth itself when dyed in indigo. In China from the third century AD coloured paper also became popular for aesthetic reasons.[43] In the Far East indigo-coloured paper, inscribed in gold or silver, was, like the parchment of Byzantium, reserved for special religious texts, often just the frontispiece, and royal commands. Examples dating from the ninth to the fifteenth centuries include Chinese and Korean Buddhist *sutras* copied in silver and gold calligraphy onto indigo-coloured paper scrolls or concertina booklets, and Chinese and Tibetan manuscripts in the Dunhuang collections.[44] Sometimes the coloured paper was glazed, as cloth still is today in China, which turned the paper a lustrous midnight blue. Analysis has shown rice powder and gypsum to be glazing substances for such manuscripts, while burnishing tools could have been made of agate, as they still are in traditional Chinese paper manufacture today at Khotan on the edge of theTakla Makan desert.[45] Some modern artists and paper makers, including the Japanese Fukumoto Shihoko and the English Gillian Spires, combine gold and indigo in their exotic work.

In the Middle East and India from the fourteenth century indigo-blue and indigo-black coloured paper was being manufactured for letter-writing and medicine wrappings. It was also used for religious texts and manuscripts, including, later, Hebrew ones, as well as for marble paper.[46] Indo–Persian texts on paper dyeing describe all the shades that could be achieved by mixing indigo with other dyes, and the way paper was indigo-dyed in the sheet.[47] The influence on Europe of oriental practices was filtered via Venice, a city with an exceptionally active dyeing industry and strong trading links with the East. Although the original attraction of the blue colour may have been its ability to counteract the general tendency of paper to turn yellow, in early sixteenth-century Venice quite a craze developed among artists for blue paper, known as *carta azzura* or *carta turchina*. Once Venetian painters like Carpaccio chose to draw on blue paper, other Italian centres of art, and later Dutch and French drawing schools of the sixteenth century, too, followed the trend.[48] In England the first patent was issued in 1665 specifically for the manufacture of blue paper for sugar wrappings as well as for art paper.[49] The influence of sixteenth-century Venice also affected certain luxury books, for master printers produced special editions on *carta azzura*. This in turn influenced deluxe printing of Hebrew religious books in northern Italy and, from the mid-seventeenth to the early twentieth centuries, in Holland and central Europe. (This is not to be confused with the cheap bluish-coloured editions produced in eastern European printing houses.)[50]

In Europe common blue paper-colorants were indigo/woad and smalt (finely ground cobalt blue glass), supplemented by more fugitive dyes such as logwood, before the stronger and more light-fast Prussian Blue came to dominate in the eighteenth century.[51] To achieve a blue colour, paper was either colour-washed or dyed, as pulp or in the piece. Dyeing of paper pulp is akin to dyeing textile fibres, and in fact the paper makers sometimes called on the help of the fabric dyers. The two come together in the case of blue rag paper, the cheapest way to produce blue paper.[52] Much of this was made from recycled indigo-dyed clothing; the Dutch, whose sailors' uniforms and country clothing were predominantly indigo-dyed, excelled in its manufacture. Today blue denim offcuts are still recycled into rag paper and also used in the manufacture of banknotes. Perhaps in the future more old clothing could be turned into paper (and money!).

السلطان
فتحعلی شاه
قاجار

Nilini [indigo] is purgative . . .
It cures afflictions by evil spirits, poisoning,
splenic disorders and upward movement of the wind.
(TIBETAN MATERIA MEDICA)

'In Sickness and in Health'

Blue Beards, Blue Bodies

Indigo's use is not confined to that of textile dye or paint pigment. We have already seen that it is difficult to disentangle the practical function of indigo as a colouring substance from all the accumulated beliefs that it embodies. As well as the applications already covered, indigo has played a role in various other subsidiary ways, many related in part to its medicinal functions. The roots of a community's appreciation of indigo, handed down as oral tradition, are difficult to dig up. They have surely arisen from several sources: awe of the strange complexity and antiquity of the dye processes, indigo's apparent healing properties, and the uniqueness of its colour. Although for the sake of clarity the various unusual uses of indigo will be examined under separate headings, in reality they are often inextricably entangled.

Direct medicinal uses of indigo throughout history

Belief in the effectiveness of herbal medicines, particularly those that double as dyestuffs, may be based as much on their colour and symbolic associations as on their supposed toxicity. Faith in the healing power of colour is ancient. We must bear in mind that in the past colour was not an abstract concept but was defined by its source in nature. An Egyptian papyrus of 1550 BC listed among 'coloured' cures 'white oil, red lead, testicles of a black ass, black lizards, indigo and verdigris'.[1] The scarcity

Kashan rug (detail), showing the Persian ruler Fath Ali Shah (d.1834) with his famous black beard (depicted in dark indigo-dyed wool) made lustrous by dyeing with henna and indigo. Late nineteenth century. (See pp.226–7.)

of blue in the natural world, and the difficulty of producing it as a dye, would have enhanced its mystery.

Indigo, considered a 'cool' and 'magical' colour was widely used for cooling feverish conditions – a clear case of sympathetic medicine. In many cultures indigo has had ambiguous associations. In India, for example, it is connected with both ill omen and with infinity, as embodied in the god Krishna.[2] Equally, in the Arab world blue has been regarded as both lucky and as so inauspicious that a person would even say 'green' when he actually meant blue, in order to avoid uttering such an ill-omened word. For the Arabs of the Middle Ages blue has been aptly described as 'a kind of homoeopathic repellent'. However pleasing as a colour, wearing it might attract the 'evil eye'; children and pregnant wives who wore blue clothes therefore needed the protection afforded by blue beads and amulets whose function was to deflect the 'eye'.[3] The protective function of blue amulets has retained its potency in much of the Arab world, where various antidotes to death and illness, such as burning blue cloth and paper, have been recorded.[4] Similarly in Indonesia textile colours have embodied strong links with life's cycles, including death and regeneration. On the island of Sumba, according to Hoskins, 'a body of occult knowledge known as *moro* (blueness)' expounds indigo's place in herbal medicines and secret rituals, including witchcraft. A woman indigo dyer, a 'person who applies blueness', may also qualify as a midwife and herbal healer.[5]

So much for the power of colour in dyestuffs – what about their medical potency? At present pharmacists are not able to explain *why* the chemicals present in indigo and various other dyestuffs seem to have some genuine medicinal effect but accept that this may be so. Although indigo is not among the world's most notable medicinal plants, the fact that it has featured in traditional medicine is not so surprising when one examines their function in nature. Chemical constituents are not present in certain dye plants and dye insects in order to furnish mankind with colouring matter, but may well be there to protect the plant or insect and repel predators. In some plants where the coloured pigments are obvious, it is possible that the colour itself attracts insects or birds for the purpose of pollination.[6] The function of a colourless plant component like indican, indigo's precursor, which requires a chemical conversion to become a coloured dye, remains a matter for speculation. It is, though, perfectly feasible that the chemicals in dye plants and dye insects that man has exploited for their colour should also have medicinal properties which in future will be better understood.

Throughout history the list of ailments indigo has been reputed to cure is extensive, mostly based on its apparent antiseptic, astringent and

Illustration of woad in a fifteenth century Persian manuscript of Dioscorides' *Materia Medica*.

Page from Salmon's *English Herbal* of 1710 describing various concoctions made from woad and listing their many medicinal uses.

1272 Salmon's *Herbal*. I

The Virtues.

X. *The Liquid Juice.* It conſolidates Green Wounds, uniting their Lips ſpeedily together; and taken inwardly 2 or 3 Spoonfuls at a time in Wine and Water, it ſtops inward Fluxes of Blood, and Cures inward Wounds: it ſtops the overflowing of the Terms in Women, Cures Spitting and Vomiting of Blood, the Hepatick Flux, Bloody Flux, and all other Fluxes of the Bowels. It is ſaid to Cure Ulcers and Wounds in the Reins and Bladder, Womb, and other ſecret parts, as alſo Ulcers and Fiſtula's in any other part of the Body, being inwardly taken and outwardly applyed: not being inferior to *Agrimony*, *Avens*, *Betony*, *Burnet*, *Comfry*, *Daiſies*, *Golden Rod*, *Horſetail*, *Knotgraſs*, *Ladies Mantle*, *Mouſe Ear*, *Madder Roots*, *Periwinkle*, *Sanicle*, *Tormentil*, or other Herbs of like kind.

XI. *The Decoction in Wine and Water.* It has all the former Virtues, but not altogether ſo powerful; and may be given Morning and Night, from 3 Ounces to 6, ſweetned with Syrup of the Juice of the ſame. It heals inward Ulcers in the Reins and Bladder, and hinders Inflamations, being fomented upon any part affected.

XII. *The Balſam or Ointment.* It is made with Hogs Lard, or with Oil Olive, Bees Wax, and a little Turpentine. It heals all manner of Wounds, and Sores: the *Germans* uſe it very much, and extol it beyond any other Balſam made of a ſimple Herb. It is no leſs helpful for foul Ulcers and Fiſtula's, hard to be Cured in what part of the Body ſoever, and heals Cankers of the Mouth and Gums.

XIII. *The Cataplaſm of the Green Herb.* It is Aſtringent and Glutinous withal, and a ſingular remedy to be preſently applyed to ſimple Green Wounds, to conſolidate them. It alſo gives eaſe in the Gout, and abates the Tumor.

XIV. *The Diſtilled Water.* It has all the Virtues of the Juice and Decoction, but much inferior in Virtues and Effects; and therefore may be uſed as a Vehicle to convey the other Medicines down in. But it may be given of it ſelf for the ſame purpoſes from 4 to 6 Ounces, ſweetned with Syrup of Comfrey, or Syrup of ſome of the other Conſolidatives.

XV. *The Pouder of the Herb and Flowers.* It may be given for all the ſame purpoſes in the Decoction, Diſtilled Water, or ſome proper Syrup, or in Honey, from 2 Scruples to a Dram, or Dram and half, Morning and Night.

CHAP. DCCXL.

Of WOAD, *Garden* and *Wild.*

I. THE *Names.* It is called in *Greek*, Ἰσάτις: in *Latin*, *Glaſtum* and *Iſatis* alſo; and by ſome *Guadum*: and in *Engliſh*, Woad.

II. *The Kinds.* We have but two Kinds hereof, *viz.* 1. Ἰσάτις ἥμερος: *Glaſtum ſativum*, *Iſatis ſativa*: Our Manured Woad. 2. Ἰσάτις ἀγρία: *Glaſtum ſylveſtre*, *Iſatis agria*: Wild Woad.

The Deſcriptions

III. The Firſt, or our Common M

Its Root is white and long, growing

Woad Garde

Woad Gard

purgative properties, although it was also said to prevent or cure nervous afflictions such as hysteria, epilepsy and depression. All parts of the plant have been used – crushed leaves, dye pigment, roots and seeds. They have been mixed into unlikely concoctions with all sorts of substances including goat hair, egg white, castor oil, fat, polenta and pepper, and applied externally as well as consumed orally. Such uses are recorded before the first centuries AD in India and China[7] as well as in Ancient Greece and Rome. Hippocrates in the fourth century BC suggests a woad treatment for ulcers, and in the first century AD Pliny[8] and Galen refer to woad's healing powers, as does Dioscorides, who also mentions indigo in his seminal *Materia Medica*.[9] The early Arabic *Materia Medica*, which preserved and overtook Greek and Latin antecedents, frequently emphasize the antiseptic and other medicinal uses for indigo. Arab physicians, with their holistic approach to medicine, were expected to have a familiarity with philosophy, dietetics, mathematics and pharmacy. The thirteenth-century Arab botanist, Ibn al-Baytar, in his great pharmacopoeia known as the *Traité des Simples* (which included observations of Dioscorides, Avicenna and al-Razi), lists a long string of uses for indigo, many of them based on its cooling qualities.[10]

From the eleventh century much of the Islamic corpus of scientific literature was transmitted to Europe through Islamic Spain and then spread via early printed books. The resulting European interest in medicinal plants was reflected in monastic gardens. European herbalists and physicians often copied their Classical predecessors, as did de l'Obels in his *Stirpium Historia* of 1576 when informing us in lurid terms that woad leaves check haemorrhages, attacks of St Anthony's Fire, gangrene, and putrid ulcers.[11] Salmon's *English Herbal* of 1710 tells us that woad juice, being 'binding and very cloying',

resists putrefaction, stops Bleedings of all sorts, whether inward or outward, by the Mouth, Nose, Fundament, or private Parts; and therefore is profitable to stop the overflowing of the Terms and Loches in Women: used to Green Wounds, it sodders up their lips and quickly heals them; and is no less profitable to cleanse and correct the putridity and malignity of old running Sores, and eating Ulcers, rebellious Fistulas, pernicious Cancers, and the like . . .[12]

Woad powder, meanwhile, 'is good against the Bloody-Flux, as also all other Fluxes of the Belly, or Defluxions of Humors upon any part, vehement Catarrhs, and the like'.

Salmon, like Gerarde before him,[13] also recommends taking woad juice internally, but the benefits of doing so were disputed by herbalists such as Culpeper.[14]

Given the apparent chemical potency of indigo it is not surprising that it would be toxic if taken internally in excessive amounts, and recent scientific research has confirmed the toxicity of some indigofera species.[15] Certainly forced labourers on the indigo plantations of East and West suffered ill-effects, even fatalities, due to processing indigo. The toxicity of small quantities taken internally may have been beneficial according to homoeopathic principles, but in the nineteenth century doctors were still at odds about its effects on the system, for we learn from one source that: 'Some physicians recommend indigo in the quantity of a dram, while others condemn the practice, and look on it as a poison. The internal use of indigo is prohibited by law in Saxony.'[16] However, its leaf juice was used in China as an antidote to poisoning,[17] and one Chinese source, which lists many medicinal uses of indigo, includes a recommendation that women should drink indigo leaf juice after childbirth, but warns that an overdose can be 'clearly injurious'.[18] In sub-Saharan Africa, where medicine men made widespread use of indigo, including decoctions of the root bark to eliminate worms among the Zulus, one herbalist did indeed accidentally cause the death of a woman by administering an overdose of indigo.[19]

Indigo dyer who had bandaged a leg wound with an indigo rag as he was convinced it was a more effective antiseptic than modern ointments. Zabid, Yemen, 1989.

Until the present day the use of indigo to treat burns, bites from insects, snakes and animals, intestinal worms, fever and stomach disorders, has continued, especially in the Far East, India, the Middle East and parts of Africa.[20] One Indian source also recommends indigo treatments specifically for rabies and cholera,[21] while in Turkey a patient would be held directly over an indigo fermentation vat as a remedy for jaundice.[22] In Central and South America, where indigo's medicinal qualities were appreciated by indigenous populations such as the Aztecs long before the Spanish conquest, a Mexican compendium of sources from the sixteenth to the twentieth centuries agrees that indigo could be used as a purgative, and to calm all kinds of stomach disorders. It could also, as the Chinese believed, control nervous disorders, particularly epilepsy, and the Guatemalan Indians even bathed their domestic animals with indigo solutions to control mites and other insects.[23]

In countries in the southern Arabian peninsula in the late 1980s many women in isolated communities were still anointing themselves and their offspring with extracts of wild or commercial plant indigo.[24] For this reason dyers were selling dyestuff alongside dyed cloth. Men, too, were applying indigo treatments, sometimes choosing them in preference to available modern medicine. Some, for example, remained convinced that wrapping an indigo cloth coated in beeswax and oil around a wound was more effective than modern antiseptic lotions. Among the Omani bedouin indigo was such a useful panacea it earned the nickname *haras*, 'the guardian'. Underclothing was even being dyed in indigo specifically to wear next to the skin to propitiate the *zar* (demanding pre-Islamic evil spirits).[25] Indigo paste was rubbed onto a child's body to cool a fever and smeared onto a new born baby's navel and around its eyes 'for good luck'. This could have been a sound prophylactic precaution, since indigo has been used since Classical times to soothe eye ailments, and recent Chinese research has indeed shown indigo root extract to be an effective treatment for conjunctivitis and trachoma.[26]

Women in indigo-dyed skirts. The skirt adorned with coloured embroidery indicates the status of the wearer as a medicine woman (including a performer of female circumcision). Dogon village, Mali, 1997.

Beliefs linking indigo with female fertility in many cultures have already been discussed (see p.126). It was used as an antidote to various sexually transmitted diseases by the Aztecs and other peoples of Central America,[27] in India,[28] and in East Africa where such ailments were treated with the root of *Indigofera tinctoria*.[29] In China its roots and stalks were recommended for menstrual problems.[30] In the former indigo-growing oases of Upper Egypt a few men were still growing indigo in the late 1980s to provide juice for traditional fertility rituals.[31] Also in Egypt Walker recorded, in the early years of the twentieth century, a belief that if a woman wearing an indigo dress came near a woman who was giving birth, the latter would lose her ability to conceive in the future; however, this affliction could be 'cured' if she paid a visit to an indigo dye works, the experience being known as *Mushahhara bi'l-Nila* ('blackening with indigo').[32] Among the Hausa of Nigeria, as among the Mexicans,[33] indigo extract was recommended as a contraceptive or abortifacient.[34]

A similar linking of indigo with fertility has been common in parts of Southeast Asia, where, as has been described, the presence of a fertile woman was considered detrimental to the dye bath or, conversely, the dye bath could cause a spontaneous abortion.[35] On the Indonesian island of Sumba, to quote Hoskins again, 'the art of traditional [indigo] dyeing is merged with the production of herbal medicines, poisons, abortifacients and fertility potions'.[36] Indeed, the same ingredients as were needed to keep the dye vat healthy were also administered to a woman after childbirth to control bleeding.

In Nigeria even horses and donkeys were treated for stomach problems with a solution of indigo administered via the nostrils.[37] An Indian dyer from Jaiselmeer in the desert region of Rajasthan in the early 1990s had a bag of large tablets made from indigo sediment collected from the base of his dye vat. These were said to help keep camels cool when administered during the hot season.[38]

Recent scientific research is beginning to lend credence to many historical claims for indigo's medical efficacy.[39] In China, where woad medicine is routinely prescribed for viral and bacterial infections, new

A baby dressed in an indigo-dyed tunic as this was believed to be protective and antiseptic. Wadi Daffa, Saudi Arabia, 1984.

trials suggest that it may be effective in treating pulmonary complications associated with cystic fibrosis.[40] In other tests using pigment and infusions from all the indigo plants, mumps, hepatitis, eczema, chickenpox and meningitis have been successfully treated.[41] Furthermore, indirubin, present in varying degrees in plant indigos, can help to combat certain cancers.[42]

Indigo-dyed clothing as prophylactic

Beliefs in the antiseptic and repellent properties of indigo have extended to the wearing of indigo-dyed clothing for protection. This is the case in many countries and may partly relate to the lingering ammonia-like smell which emanates from newly dyed indigo cloth, as well as to indigo's healing powers and sometimes to the other qualities associated with it already discussed. The surface rub-off from new cloth, an undesirable property to the Western way of thinking, has been considered a positive asset in other cultures, both for medical and for status reasons.

Farming communities of China, Japan and other parts of Asia generally held that indigo-dyed clothing would repel snakes and insects lurking in the rice paddies.[43] In rural Japan in the 1950s washing lines festooned with indigo-blue babies' nappies were still a common sight. In the 1980s some Yemeni babies were still being dressed in indigo-dyed tunics, their navels wrapped round with an extra piece of indigo-dyed cloth. The main attraction of such indigo-dyed cloth was its supposed gentle antiseptic qualities when worn against a child's delicate skin. In the Persian Gulf region, when a woman made a face mask from indigo-dyed cloth, she would keep the remnants handy for rubbing onto her children's wounds when required.

The dramatic indigo-dyed turban also had its medicinal aspect. It was thought to prevent and relieve headaches, as well as protecting the wearer from djinns. Headaches were also relieved by inhaling smoke from burning indigofera roots.[44] For centuries the blue robes, and especially the shiny indigo-dyed turban, of the north-west Saharan 'blue men' – the Reguibat – as well as the Touareg and some other tribal groupings of the Sahara and Arabia, have protected them from the harshness of the desert climate.[45] The blue staining of the hands and face that resulted both had cosmetic value and became a symbol of tribal identity. Such beliefs tend to linger on in a nebulous way; young men wearing turbans dyed with synthetic indigo in southern Morocco for example still claim that they afford protection 'from everything'. Similarly, for many Middle Eastern women the shiny indigo-dyed mask was considered generally protective, the consequent rub-off of dye onto the skin being positively welcomed [46]

Cartoon by Arnold Roth in *Punch* magazine showing the Ancient Britons applying woad paint to frighten off the invading Romans. In the cartoon it starts to rain, which washes off the paint; the Romans, no longer alarmed, thus successfully invade Britain.

Cosmetics

Closely linked with medicine is indigo's application as bodily adornment, whether for cosmetics, for tattooing or for dyeing head and facial hair. Here again the aesthetically desirable was underpinned by both practical and mystical beliefs.

Whether or not the Ancient Britons frightened their enemies by covering themselves in woad paint may remain uncertain,[47] but as the invading Romans dubbed them *Picti*, i.e. 'painted men', they probably used some coloured substance as facial or body paint or even for tattoos. As already noted, we do now have archaeo-botanical evidence of the existence of woad in Britain by the time of Julius Caesar,[48] and one day chemical analysis of skin remains from ancient burials may settle the matter one way or the other. As an ironic illustration of the circularity of history, a new woad product is now becoming chic in France, where top cosmetic houses are trying out woad seed oil to produce an 'eco-friendly' cosmetic base.

In other parts of the world indigo has frequently been rubbed on the body, for decoration, for medicinal reasons or for a combination of the two, as in tattooing. In Central America, for example, indigo was among the colours employed by the Mayas and Aztecs to beautify both face and body. While black was used there by unmarried men and red by warriors, blue was reserved for the priestly caste and those about to be sacrificed.[49] The practice was discouraged by the church after the Spanish arrived.

In most places the crushed pigment was mixed with a chalky substance, but women in central Africa seem to have gone to peculiar extremes to manufacture indigo body paint. Captain Clapperton, travelling in northern Nigeria in the early nineteenth century recorded:

The women of this country, and of Bornou, dye their hair blue as well as their hands, feet, legs and eyebrows. They prefer the paint called shunee *[= prepared indigo], made in the following manner:-They have an old tobe [indigo-dyed robe] slit up, and dyed a second time. They make a pit in the ground, moistening it with water, in which they put the old tobe, first imbedded in sheep's dung, and well drenched with water, and then fill up the pit with wet earth . . . After seven or eight days the remnants of the old tobe, so decayed in texture as barely to hang together, are taken out and dried in the sun for use. This paint sells at 400 cowries the gubga, or fathom [six feet]; for this measure commonly gives its name to the cloth itself. A little of the paint being mixed with water in a shell, with a feather in one hand, and a looking-glass in the other, the lady carefully embellishes her sable charms. The arms and legs, when painted, look as if covered with dark blue gloves and boots.*[50]

In many parts of the Arab world where both men and women have considered indigo decorative and good for the complexion, it was used for tingeing the face blue for special occasions such as festivals and weddings.[51] Even in the 1990s it was being applied by women in Yemen, and among the Touareg of the Sahara, for this purpose. It was also applied in decorative patterns to the face, hands and feet, like henna, and indeed was sometimes called 'black henna' when the two herbs were combined to make a dye also used for the hair (see p.226). Freya Stark refers to a young bride of the Hadhramaut with her hands 'done delicately in an intricate pattern of blue henna'.[52] A trademark of the Yemeni tribesman was the habit of smearing the body with indigo mixed with an emollient such as sesame oil both to provide protection and as a symbol of virility. Most travellers to the Arabian peninsula over the centuries, including the fifteenth-century Chinese navigator Zheng He and, in the twentieth century, Wilfred Thesiger, were struckby the indigo-stained skins of the tribesmen.[53] Freya Stark, describing an encounter with some Wahidi tribesmen, enthused: 'Their beauty was in the bare torso, the muscles rippling in freedom under a skin to which a perpetual treatment of indigo, sun and oil gives a bloom neither brown nor blue, but something like a dark plum.'[54]

Woman of the Jiddat Harasis tribe wearing a protective indigo-dyed mask, whose dye rub-off onto the skin was considered beneficial to the complexion. Between Hayma and Adam, Oman, 1988.

Tattoos

Indigo has proved an ideal substance for staining tattoos, given its supposed antiseptic qualities and taking into account the symbolism of the colour blue. The origins of tattooing and scarification, which span the continents, are obscure, but may stem from a primarily medical function dating back to prehistory. Evidence exists of tattooing in prehistoric America, and in Egypt the disembalmed body of a woman of Thebes who lived 5000 years ago shows traces of

Bedouin woman grinding up indigo and applying the paste as make-up in preparation for a festival. Sayhut, Yemen, 1990.

scarification, stained white and blue, on her abdomen; its configuration suggests a medical, rather than an ornamental, purpose.[55] The Copts and Muslims, but particularly the former, continued to use tattooing in medicine. Children were tattooed as a prophylactic precaution (e.g. on the temples against migraine) and to cure all sorts of afflictions. Although other plants with antiseptic qualities were also used, indigo was popular,[56] particularly in poorer areas, as recorded by one traveller in Egypt at the end of the nineteenth century, who noted that women had tattoos on hands, arms, feet, cleavage, forehead, chin and lips.[57]

Tattooing with indigo was common in other parts of Arabia and Africa too. A 1946 British wartime intelligence handbook on Western Arabia tells us that: 'Nearly all women are tattooed with indigo on lips, cheeks, nose, breast and abdomen, generally in a pattern of circles and triangles.'[58] In the 1980s and 1990s some bedouin of both sexes still had blue tattoos. As an example of the special status accorded to tattooists, Moroccan women practitioners were blessed by the local saint or more dubious spirits, and the occasion of a tattooing was a cause for celebration. Tattoo designs, which, as elsewhere,[59] often resembled those found on tribal weavings, had specific functions: some served as preventative medicine while others were talismanic, acting to deflect the evil eye.[60]

Further south in the sub-Sahara it was, for obvious reasons, those with lighter skins who could proudly display indigo-dyed tattoos, while cicatrization, which produced permanent raised scars, was used by darker-skinned people. Some people, though, like the lighter-skinned Sokoto Fulani of north-western Nigeria, deliberately rubbed indigo into their facial wounds to produce blue scarification.[61]

For centuries tattooing was also practised in the East. Indigo tattooing may have been practised in northern Thailand as it was by one minority group of southern China,[62] as a vestige of a much older tradition. Ralph Fitch, an English merchant travelling in Burma in Shakespeare's time, noted that indigo-blue tattoos were a mark of the aristocracy:

The Bramas . . . have their legs or bellies, or some part of their body, as they thinke good themselves, made black with certaine things which they have; they use to pricke the skinne, and to put on it a kinde of anile [indigo] or blacking, which doth continue alwayes. And this is counted an honour among them; but none may have it but the Bramas which are of the kings kindred.[63]

Hair and beards

Once again the colour and lustre of indigo combined with its apparent medical properties have given it a dual attraction. It is not known which people first dyed their hair with woad or indigo but it is has been suggested that Ovid, who mentioned the practice, may have been referring to the early Germanic Teutons.[64] Whatever the original reasons for adopting indigo hair dye it has featured in the ritual uses of hairdressing to transform the appearance and indicate social status. A priest in Mexico in 1547 noted the custom among indigenous women of dyeing their hair with black clay or indigo to make it shine.[65] Indigo is still used for this purpose in parts of Africa and southern Asia.

In early Islam the custom of dyeing hair and beards with indigo and henna was a much debated issue,[66] but the practice has continued to this day in a culture where grey hair is not much appreciated and where facial hair is a significant symbol of masculinity and religious pride. (Could this even explain the name of the legendary folk hero Bluebeard?) In the early eleventh century Al-Biruni noted that indigo 'masks the defects of ageing' and 'bestows *samat* (beauty)'.[67] Persian men have always been particularly partial to black facial hair, but noblewomen, too, blackened their eyebrows with indigo, as did women of China and Japan. The European traveller Chardin in the late seventeenth century remarks:

Indigo-based products still produced in Pakistan and India. The Black Vasma Henna is a henna/indigo mixture used as a hair dye (*wasma* is Arabic for indigo leaves) and Anoop is an Ayurvedic hair tonic. Also dried indigofera leaves and indigo pigment.

Black hair is most in Esteem with the Persians, as well as the Hair of the Head, as the Eye-brows and Beard: The thickest and largest Eye-brows are accounted the finest, especially when they are so large that they touch each other. The Arabian Women have the finest Eye-brows of this kind. Those of the Persian Women, who have not Hair of that Colour, dye and rub it over with Black to improve it ... They likewise generally annoint their Hands and Feet with that Orange-colour'd Pomatom, which they call Hanna, which is made with the Seed or Leaves of Woad or Pastel ground ... which they make use of to preserve the Skin against the heat of the Weather.[68]

Here is a clear case of confusion between the roles of henna, woad and indigo. Indigo and henna were in fact often combined to make hair dye sometimes called simply 'henna'.

In the early nineteenth century another European traveller to Persia noted that, despite much indigo being imported from India, it was also widely cultivated in southern Persia for dyeing both linen and beards.[69] Black beards were particularly in vogue among the aristocracy at this time in emulation of the ruler, Fath Ali Shah, who modelled his own appearance on predecessors whose fine beards are portrayed in ancient sculptures. The famous beard dominates innumerable depictions of this Shah in European-

style oil paintings and pictorial rugs popular in nineteenth-century Persia. We learn how he achieved the required effect from a vivid description of the lengthy activities of the heir apparent in the royal bathhouse at Tabriz:

[The attendant, having washed the bather] takes his employer's head upon his knees, and rubs in with all his might, a sort of wet paste of henna plant, into the mustachios and beard. In a few minutes this pomade dyes them a bright red . . . The next process seizes the hair of the face, whence the henna is cleaned away, and replaced by another paste, called rang, *composed of the leaves of the indigo plant. To this succeeds the shampooing . . .*[70]

The indigo and henna mixture was also popular amongst Muslims of the Punjab,[71] and by the late 1930s some places in India were cultivating the plants solely for hair dye and treatment.[72] In neighbouring Pakistani Sind indigo is also grown for hair dye used locally and exported as 'black henna' to the Middle East and elsewhere.[73] The product was even found on the shelves of an Indian shop in a small town in Belgium by this writer in 1997, labelled 'Black Vasma Henna', 'wasma' being the name used in medieval Islamic literature for indigo leaves.

In Nigeria Hausa women have applied indigo to the hair,[74] but for the Yoruba in particular hairdressing has played an important part in traditional culture. Here indigo featured in palace life in various hairstyles and rituals signifying social hierarchies. A court messenger, *ilari*, had his hair ritually shaved in one of the various distinctive patterns, leaving circular patches of long hair to be braided and dyed in indigo. Court messengers' hairstyles were still prominent in some Yoruba palaces in the 1970s.[75] Among the Yoruba too, devotees of the important trickster god Eshu, whose wooden images were often stained with indigo, would darken their beards with indigo and wear contrasting white cowrie shells as a visual metaphor for the ambivalent status of the god.[76] In the Solomon Islands even head-hunters' ritual wigs, made from hair taken from the dead, were sometimes dyed blue.[77]

On purely medical grounds, numerous sources in Arabia and the East recommend indigo in a variety of forms for hair and scalp problems. According to Ibn al-Baytar in the thirteenth century, baldness and head ulcers could be prevented with indigo treatments.[78] Dandruff, head lice and scalp itches were treated with ash from the burnt plant,[79] or a decoction of powdered root or seed (in one instance infused in rum!)[80] or a leaf extract.[81] In the Malay Peninsula a poultice of indigo leaves placed on a child's head was thought to draw out worms.[82] Even today an extract of *Indigofera tinctoria* is a main ingredient of a popular Indian Ayurvedic herbal hair oil, which claims to 'tone up both scalp and hair and arrest hair fall'.[83]

chapter ten

Into the Future

Like the rainbow in which it features, indigo has always formed an arc across societies in every direction and continues to do so today. Blue jeans, whose colour is of Eastern origin, have taken indigo from the West back into the East, linking cultures and generations. Meanwhile, the attraction of ethnic textiles and handmade crafts brings elements of the so-called 'Third World' into the 'First'. Indigo, for so long the main dye in many rural societies of the 'Third World', has resurfaced in sophisticated urban high streets, often in the guise of Asian batiks and African prints. In the Western home oriental carpets, with their blues and reds, are everywhere. As a sartorial statement blue and white has an enduring freshness, especially for summer clothing, while indigo has both a classic and a nostalgic appeal. Thanks to the growing preoccupation in the West with natural products, indigo may even regain its popularity in countries where, though once the norm, it is now out of favour. The manufacture in Mali of 'mud-cloth' – a natural product if ever there was one – has enjoyed a dramatic reversal in its fortunes since the West discovered its attractions. Indigo may by a similar process recover lost ground.

Already, despite its generally undeserved aura of ecological correctness, indigo is being linked with a revolution in the manufacture of products, which aims to maximize the use of renewable natural resources such as linseed straw and hemp. Indigo and bast fibres have been allies since

Rainbow with the mysterious indigo hovering between violet and blue colours. Some people are unable to distinguish the indigo colour.

antiquity, and the world's small-scale revival of natural indigo complements this renewed interest in natural fibres. They are ideally suited for weaving into hard-wearing cloth, which almost replicates the original jeans worn in the heady days of the gold rush in California. Could scientific advances in genetic engineering lead before long to strange new ways of producing 'natural' indigo that could realign the dye closer to its biological roots? Needless to say, the word 'natural has more than one connotation.

Although the worn look of faded jeans is already appreciated, we could go much further in truly valuing used clothing, not just for reasons of economy, but for its own sake. The Japanese, for example, took pride in refashioning a kimono first into a short coat, then into sleeves for another garment, followed by use as part of a domestic furnishing, until finally the scraps functioned as household rags or were torn up and re-woven into new textiles.[1] Used indigo-dyed cloth and garments and left-over factory-dyed yarns and scraps are tailor-made for this kind of treatment, as a few Western textile manufacturers and fashion designers have already realized. In the West, however, a true appreciation of the riches to be found in rags has yet to hit the high street, with its vested interest in consumer spending and its extravagant obsession with novelty for novelty's sake. But, after all, in the world of fashion ideas are continually being recycled, so why not textiles themselves? If this happened the 'Rag Trade' (as the textile industry terms itself with false modesty) could really take on a more literal meaning.

Japanese country jacket, made, like a rag rug, by weaving strips of used indigo-dyed cloth. The neck area is reinforced with hemp quilting.

In Britain until after the Second World War scraps of worn woollen government uniforms (police, navy, etc.) were turned into 'shoddy', which was reused by being mixed with dyed virgin wool to eke it out. Although this did not apply solely to indigo-dyed cloth, the activity stopped once government uniforms ceased to be dyed mainly with indigo blue. In

France and Spain some denim is indeed reclaimed, despite problems with the flammability of cotton.[2] But if the demand was created, there would be scope for more widespread recycling of textile fibres, reusing their colour and at the same time benefiting the environment. Indigo-dyed fibres are particularly suitable for such treatment.

And how about reusing not just clothing but dyes themselves? Here again we could also learn from the past to be less profligate and save resources. The Dutch, Spanish and others were recycling dyes, especially the expensive reds but also blues, in the seventeenth century if not before, in their case in deference to their pockets rather than to the environment. It was accepted practice at that time to remove the more valuable dyes from wool shearings, or from the left-over flock on the teasels used to raise the nap of woollen cloth, and to reuse it to dye stockings, aprons and cheap blankets. Even in the nineteenth century blue wools for Transylvanian rugs were dyed with the indigo from commercial felt cuttings.[3] The method used was to extract the indigo dye in successive baths of a strongly alkaline solution and then to re-vat the dye. The main discouragement today is that, in contrast with the past, re-extracting indigo costs more than buying new dyestuff.

What in any case are the environmental implications of indigo dyeing – and is natural dye better than synthetic? Or have both had their day? And are entirely new industrial battlefields looming?

As regards natural plant indigo, this can certainly be a useful alternative crop for supplying a niche market. We have already seen how climate changes may open up new possibilities for growing indigo or woad on surplus ground in places which may not have been suitable in the past. And we have seen how this is bringing the ancient woad story up to the minute for use in the cut-throat world of commerce as imaginative ideas for marrying natural indigo with modern technology are emerging. Woad need not deplete the soil if properly managed, while indigoferas provide excellent green manure from their waste products when used for dyeing. However, wide-scale cultivation of indigo plants requires far too much land to be globally viable in an over-populated world. With the annual worldwide consumption of synthetic indigo running at around 20,000 tons, several million acres would be required to manufacture the equivalent by traditional methods.[4]

If plant indigo is too greedy for land, does this signal good news in perpetuity for the manufacturers of synthetic indigo? Not necessarily. The petrochemical industry, of which aniline is a by-product, will not last forever, whereas vegetable sources are infinitely renewable, and non-toxic. Although the manufacture of indigo colorant is reasonably safe, the chemicals involved, and their by-products, are still environmentally undesirable, and alternatives are always being sought.

So what other possibilities are lurking around the corner? Genetic manipulation, whatever misgivings may be entertained about the general concept, could take the indigo story in several parallel directions. Firstly, it may be possible to use an indigo gene to modify closely related species and produce, for example, new indigo-bearing strains of *Polygonum* or *Indigofera* that could tolerate cooler climates. Secondly, the Genecor company is already making bio-indigo by genetic engineering using the standard micro-organism E-coli.[5] (Initial problems when indirubin, the reddish 'impurity' of the plant product, also appeared, were solved by splicing in another gene to remove this colour.) If this advance in bio-technology takes a commercial hold, it will be the first industrial chemical produced this way. Indigo could, therefore, before long be 'growing' once again, but this time in a bacterial form fed on glucose in a city laboratory, bypassing the countryside altogether and without using toxic chemicals.

Until bigger changes follow, bio-indigo will still be used as a dye in the modern way. Traditional fermentation vats are the most environmentally sound, and the logical option for use with any kind of natural indigo, but unfortunately they are time-consuming and therefore expensive. However, the recent isolation, by researchers at England's Reading University, of the previously unknown bacterium (a strain of *Clostridium*) responsible for making an indigo vat 'work', is a crucial breakthrough.[6] It should lead to a viable industrial bio-chemical fermentation vat to replace the current inorganic vats which are efficient but employ polluting dithionites as reducing agents. For the moment inorganic vats are best used continuously to minimize harmful waste products. Other encouraging developments are moves towards pre-reduced indigo, already being marketed by BASF, and the current research into electrochemical reduction methods using indirect electrolysis.[7] Indigo residue can at least be recovered and re-vatted, while waste rinsing water, which has a high Ph caused by the caustic soda, can be easily neutralized. The filtered floccules create a blue residue which, like textile factory spoils themselves, are suitable for converting into paper.[8] When indigo sludge is incinerated it can cause acid rain, but used as land fill or fertilizer it is kinder to the environment. Being a cool dye, and one that is developed by air, indigo does at least save on certain energy costs, and 20 per cent less dyestuff than other blues is required to produce a dark colour on a given weight of cloth.[9]

The third genetic possibility, surreal as it sounds, is to 'grow' indigo invisibly (who said indigo wasn't magical?) and side-step the dyeing process altogether. This idea has ancient roots in indigenous farming communities of central South America. Five thousand years ago farmers there had already begun to deliberately select naturally pigmented cotton varieties for their colour. Brown cotton fishing nets, for example, were invisible in water. Building on past experience, Sally Fox in California pioneered the conservation and development of organically coloured cottons in the USA in the 1980s, in order to supply such major companies as Levi Strauss and Co.[10] Until recently, though, conventional selective breeding techniques limited the colour range mainly to browns, greens, yellows and rusts, and the blues remained elusive. But in 1996 the first US patent was granted, to the agricultural biotechnology company Calgene, for transgenic colour alteration, the aim being to produce genetically modified cotton plants that will grow black, red and blue 'naturally', whatever that means. 'Blue gene' cotton may soon be a commercial reality.[11] Although the gene being pioneered comes from a blue flower and not an indigo plant, will this scientific leap enable denim-like yarn to be produced without dyeing, thereby upstaging the makers and users of synthetic indigo? And would this procedure be limited to cotton or in time could other fibres be similarly self-blued, perhaps eventually with an indigo gene itself? And could the popular stone-washed look of jeans still be achieved? Or would clever marketing alter the public taste? Finally, is it conceivable, in a world that has already created sheep that are 'self-shearing', that scientists may one day go even further by adding indigo genes to sheep to give them blue wool, thus adding a new meaning to the expression 'dyed-in-the-wool'?

One way or another, as in the field of medicine, it seems possible that advances in genetic engineering will provide indigo with the biggest change in its fortunes since the synthetic usurper pushed plant indigo off its throne. Strangely, key developments in indigo's long reign will then have occurred about a hundred years apart, on either cusp of the twentieth century. Like a monarch keeping up with the times, indigo will surely adapt in order to hold on to its position as the 'king of dyes'.

A final word

Certain words contain within them, like Russian dolls or pass-the-parcel packages, layer upon layer of meanings as the wrapping is peeled away. I hope this book, having delved through many layers, has demonstrated that there is far more to the word 'indigo' than meets the eye.

Appendix: Chemical Formulae

The production of natural indigo and the by-product indirubin

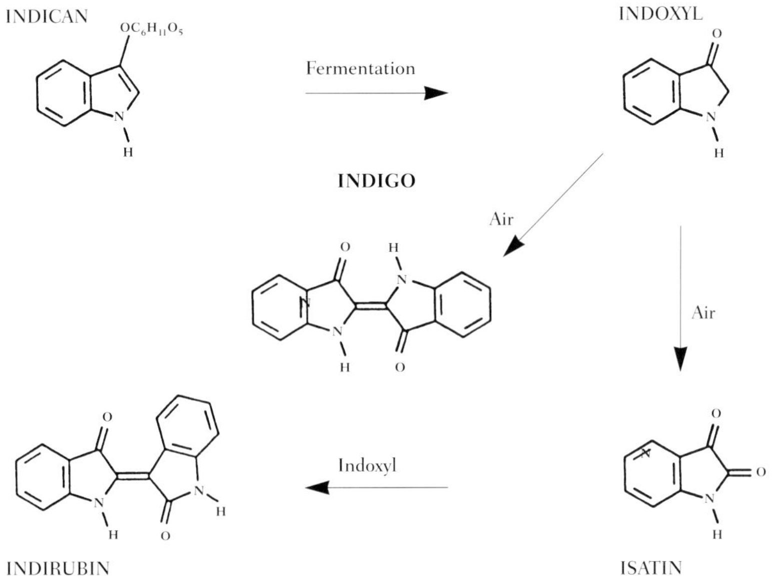

INDICAN

$OC_6H_{11}O_5$

N H

Fermentation

INDOXYL

O N H

INDIGO

Air

O H N N H O

Air

O O N H

Indoxyl

O N H O

INDIRUBIN

ISATIN

Indican, the main precursor of indigo contained in the leaves of 'tropical' indigo plants

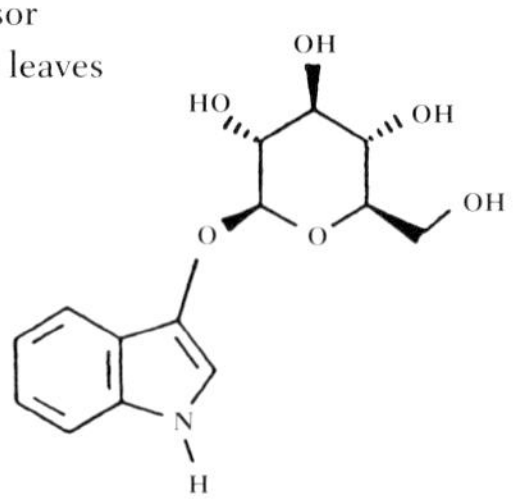

OH HO OH OH O O N H

Isatan B, a precursor of indigo contained in the leaves of *Isatis* species (woad)

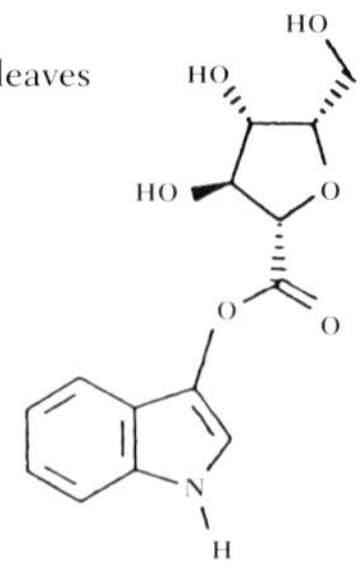

HO HO HO O O O N H

The reaction involved in indigo vat dyeing

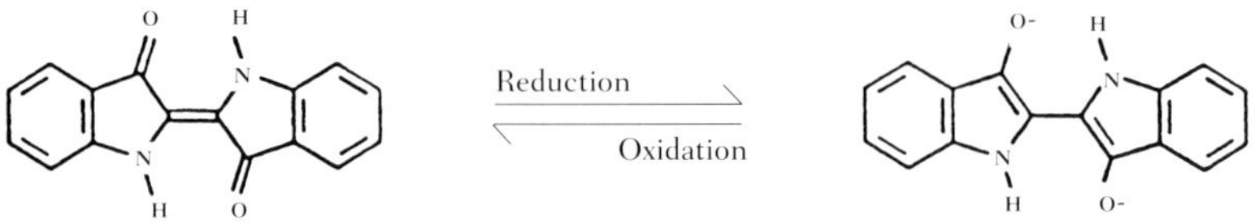

O H N N H O

Reduction

Oxidation

O- H N N H O-

INDIGO

leuco– **INDIGO**

6,6' - Dibromoindigotin – the main colourant of shellfish purple

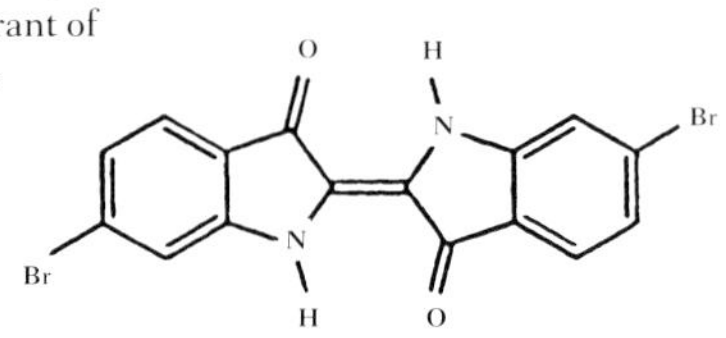

O H N Br Br N H O

Notes

Chapter One

Opening quote: attributed to William Morris.

1. BEMISS 1973: 105.
2. E.g. see MAXWELL 1990: 144; and NIESSEN 1985: 137.
3. DENDEL Esther Warner, 1974, 'Blue goes for down', *Natural Plant Dyeing*, Brooklyn Botanic Garden, New York: 23–8.
4. BILLINGTON Vivien, 'A Data-base of English Woadpeople', in CARDON *et al.*, in press.
5. See GRIERSON 1986, plate 8c, for an illustration of the way in which indigo extends the colour range.
6. ANON 1705: 315.
7. KNECHT, RAWSON and LOWENTHAL 1941, Vol. I: 314. See also BEMISS 1973: 16.
8. MOHANTY *et al.* 1987: 36–40.
9. GAGE John, 1993, *Colour and Culture*, London: 232; and MCLAREN K. 1986: 7.
10. See WEINER Annette B. and SCHNEIDER Jane (eds), 1989, *Cloth and Human Experience*, Washington DC.
11. FUKUMOTO Shigeki 1993, 'Japan and the Art of Dyeing – The Aesthetics of Age', *Ecotimes*, December: 13.
12. WOOD Frances, 1995, *Did Marco Polo go to China?*, London.

Chapter Two

Opening quote: 2 Chronicles 2:7.

1. WATT 1890: 390–91 discusses *nila*'s complicated etymology.
2. The present state of knowledge about natural dyes in history is discussed by Dominique Cardon in 'Polyphenols and natural dyes: a historian contemplates chemistry' in VERCAUTEREN J. *et al.* (eds), *Polyphenols 96*, conference papers, Bordeaux, France, July 1996, published by the Institut National de la Recherche Agronomique, Paris, 1998.
3. Ongoing research by Professor Claude Andary in the Faculty of Pharmacy at Montpellier University.
4. Examined by Pfister in the 1920s, currently being re-examined by Dr Gillian Vogelsang-Eastwood
5. For more detail see BALFOUR-PAUL 1997: Ch. 1.
6. See ZARINS Juris, 'The Early Utilization of Indigo along the Northern Ocean Rim', in JARRIGE C. (ed.), 1992, *South Asian Archaeology 1989*, Madison, Wisconsin: 469–83.
7. Reproduced in CHASSINAT E., 1921, *Papyrus Médical Copte*, Mémoires de l'Institut Français d'Archéologie Orientale du Caire, Tome XXXII: 210.
8. GERMER R., 1992, *Die Textilfärberei und die Verwendung Gefärbter Textilien im Alten Ägypten (Ägyptologische Abhandlungen)*, Band 53, Wiesbaden: 123, and pers. comm.
9. HALLEUX 1981: 138–41.
10. SALTZMAN 1992: 479.
11. SPANIER E. (ed.), 1987, *The Royal Purple and the Biblical Blue – The Study of Chief Rabbi Isaac Herzog and Recent Scientific Contributions*, Jerusalem.
12. Shellfish purple contains 6–monobromoindigotin and 6,6'–dibromoindigotin – see WOUTERS 1992: 17–21, for the chemical formulae. The High Performance Liquid Chromotography (HPLC) analysis technique distinguishes shellfish purple from indigo.
13. SOREK and AYALON 1993: 92–3.
14. DOUMET 1980 records his experiments, following Pliny's instructions, with Tyrian shellfish. See also SCHWEPPE 1992: 307–10.
15. *De Architectura*, Book VII, Ch. XIII.
16. CLARK *et al.* 1993: 197.
17. THOMPSON Jon, 1995, 'Shellfish Purple: The Use of *Purpura patula pansa* on the Pacific Coast of Mexico', *DHA* 13: 3–6; NUTTALL Zelia, 1909. 'A Curious Survival in Mexico of the Use of the *Purpura* Shell-fish for Dyeing', *Putnam Anniversary Volume* (New York): 368–84; and CARLSEN Robert S. and WENGER David A., 'The Dyes Used in Guatemalan Textiles: A Diachronic Approach', in SCHEVILL *et al.* 1991: 365–7.
18. Several experiments appear in *DHA* – and see the bibliography on shellfish purple compiled by COOKSEY Christopher J., in *DHA* 12, 1994: 57–65.
19. See SOREK and AYALON 1993: esp. 15–34.
20. In 1965 excavations ceased and were not resumed until 1993. Closer scientific analyses may reveal more information.
21. PARROT A., 1958, *Mission Archéologique de Mari*, Vol. 2, Paris.
22. FUJII Hideo and SAKAMOTO Kazuko, 1993, 'The Marked Characteristics of the Textiles Unearthed from the At-Tar Caves, Iraq', in EILAND Murray L. Jnr, PINNER Robert and DENNY Walter B. (eds), *Oriental Carpet and Textile Studies*, Vol. IV: 35–45.
23. PFISTER R., 1934, *Textiles du Palmyre*, Paris: 3–6, 24, 28, 38, 60. Textile fragments from Palmyra are currently being re-examined by Harold Böhmer in Istanbul.
24. Translated by Irving Finkel, Dept. of Western Asiatic Antiquities, British Museum.
25. WULFF 1966: 192.
26. See BALFOUR-PAUL 1997: 7–8 for refs.
27. Technical analysis appears in WHITING M.C., 1985, 'A report on the dyes of the Pazyryk carpet', in PINNER R. and DENNY W.B. (eds), *Oriental Carpet and Textiles Studies* I, London. See also BÖHMER H. and THOMPSON J., 1991, 'The Pazyryk carpet: a technical discussion', *Source*, New York, Vol. X, No. 4: 30–36.
28. Professor M. Seefelder tested a small sample he found at Mohenjo Daro in 1971 and found it to contain indigo, but this is not documented (pers. comm. 1995).
29. HUNTINGFORD, G. (ed.), 1980, *Periplus of the Erythraean Sea*, London: 62, 122.
30. WILSON 1993: 133.
31. WHITFIELD and FARRER 1990: 108.
32. LEIX Alfred, 1942, 'Ancient Textiles of Eastern Asia', *CR* 43: 1566–72.
33. WHITFIELD and FARRER 1990: 128–34; and STERLING BENJAMIN 1996: 66–74 and plates 2–4.
34. CHRISTIE, in press; and CRAWFURD John, 1856, *A Descriptive Dictionary of the Indian Islands and Adjacent Countries*, London: 156–7.
35. CHRISTIE, in press.
36. For botanical evidence see ARNOLD 1987.

37 REID 1995.
38 STONE-MILLER Rebecca, 1992, *To Weave for the Sun: Andean Textiles in the Museum of Fine Arts, Boston*, Boston: 115–16.
39 SAYER 1985: 51–69.
40 Illustrated in SEEFELDER 1994: 33.
41 JOYCE T.A., 1913, 'A Peruvian tapestry, probably of the 17th century', *Burlington Magazine*: 146–50.
42 For general background see JOHNSON Grace and SHARON Douglas (eds), 1994, *Cloth and Curing, Continuity and Change in Oaxaca*, San Diego Museum Papers, No. 32. See also TUROK Martha, 1996, 'De Fibras, Gusanos y Caracoles' ['Of fibers, worms and sea snails'], *Artes de Mexico – Textiles de Oaxaca*, No. 35: 62–9, 91–4.
43 ARNOLD 1987: 68–9.
44 ROUNDHILL Linda S., REENTS-BUDET Dorie, MCGOVERN Patrick and MICHEL Rudolph, 1994, 'Maya Blue: A Fresh Look at an Old Controversy', *Seventh Palenque Round Table* (The Pre-Columbian Art Research Institute, San Francisco): 253–7.
45 BRUNELLO 1973: 13–15; CARDON and du CHATENET 1990: 149.
46 WILD John-Peter, 'Central and northern Europe', in HARRIS (ed.) 1993: 66–7; and HALD 1980: 137–8.
47 WALTON P., 1994, 'Wools and dyes in northern Europe in the Iron Age', *Fasciculi Archaeologiae Historicae* (Poland) fasc. 6: 61–8.
48 HALL 1992: 21–2.
49 Woad seed pods from Dragonby, northern England, identified by Dr Alan Hall, York University.
50 *De Materia Medica*, Book II, Ch. 107.
51 PLINY, *Historia Naturalis*, Liber XXXV, Sect. XXVII (46).
52 SIMKIN C.G.F., 1968, *The Traditional Trade of Asia*, Oxford: 47.
53 *De Architectura*, Book VII, Ch. XIV, para. 2.
54 PLINY, (n.51).
55 FORBES 1956, Vol. 4: 112.
56 See BALFOUR-PAUL 1997: 10–12 and notes for further sources.
57 PETRIE W.M. Flinders, 1908, *Athribis*, London: 11, plates XIV and XXXV.
58 Fragments, from Maximianon/al-Zerga, are being examined by specialist Dominique Cardon. Dyed fragments, some patterned in resist, have also been excavated at the Graeco-Roman sites of Mons Claudianus and Mons Porphyrites, near present day Gebel Dokân. The latter site also has an inexplicable dump of murex shells. (Could these indicate a Roman fraud? It is conceivable that murex glands could have been used to enhance the colour of the purple porphyry, thus increasing its export value.)
59 ADAMS Nettie K., 1997, 'Sacred Textiles from an Ancient Nubian Temple', *Sacred and Ceremonial Textiles*, Proceeedings of the Fifth Biennial Symposium of The Textile Society of America, Chicago 1996: 259–68.
60 WOOLLEY Linda, 1989, 'Pagan, Classical, Christian', *Hali*, 48, December: 27–37.
61 Analysis by George Taylor and Penelope Walton Rogers of Textile Research Associates. Their results are published in *DHA* and elsewhere. See, for example, WALTON P. 1989, 'Dyes of the Viking Age: a summary of recent work', *DHA* 7: 14–20.
62 Ibid.: 18–19.
63 SCHWEPPE 1992: 61.
64 MCDOWELL Joan Allgrove, 'Sassanian Textiles', in HARRIS (ed.): 68–70. See also LEIX Alfred, 1942, 'The Sassanian Textiles and their Influence on the Western World', *CR* 43: 1559–65.
65 KAWAMI Trudy, 1991, 'Ancient Textiles from Shahr-i Qumis', *Hali* 59, October: 95–9.
66 BETHE 1984: 59–61.
67 STERLING BENJAMIN 1996: 75–93.
68 BOSER-SARIVAXÉVANIS 1969 and 1975.
69 BOLLAND 1991.
70 Ibid. (dyestuff analysis by J.H. Hofenk de Graff, Central Research Laboratory for Objects of Art and Science, Amsterdam). Reds used were brazilwood, cochineal and madder.
71 HAKLUYT 1965: 209.
72 For details see BALFOUR-PAUL 1997: 21–3. The most important source is PEGOLOTTI Francesco Balducci, 1936, *La Pratica della Mercatura* (ed. EVANS Allan), Massachusetts (reprint New York 1970).
73 For medieval Mediterranean sources see CARDON 1992b: 10 and 1990 (Ph.D. thesis): 434–48.
74 CARDON and du CHATENET 1990: 144.
75 MARCO POLO, 1875, *The Book of Ser Marco Polo* (trans. and ed. YULE Henry), London, Vol. 2: 363, 383, 388.
76 BALFOUR-PAUL 1997: 19–20, 44–5.
77 SERJEANT R.B., 1972, *Islamic Textiles*, Beirut: 110–11.
78 LOMBARD Maurice, 1978, *Les Textiles dans le Monde Musulman*, Paris: 134–43.
79 This is *Flemingia grahamiana* Wight and Arn. (syn. *F. rhodocarpa* Bak.), an important dye in medieval Yemen. Late seventh/early eighth century indigo and madder ikat fragments from Yemen or India have recently been discovered in southern Israel – see BAGINSKI Alisa and SHAMIR Orit, 1997, 'The Earliest Ikat', *Hali* 95, November: 86–7.
80 BAKER Patricia L., 1995, *Islamic Textiles*, London; SERJEANT R.B., 1972, *Islamic Textiles*, Beirut; and BALFOUR-PAUL 1997: 21.
81 MATHESIUS Anna, 'Byzantine Silks' in HARRIS (ed.) 1993: 75–9; and LEIX Alfred, 1942, 'Early Islamic Textiles', *CR*, 43: 1573–8.
82 For these dyes see BRUNELLO 1973, and CARDON and du CHATENET 1990.
83 VOGELSANG-EASTWOOD Gillian, 1990, *Resist Dyed Textiles from Quseir al-Qadim*, AEDTA, Paris; and her article, 'Unearthing History', *Hali* 1993, 67: 85–9. See also MACKIE Louise, 1989, 'Textiles', in KUBIAK W. and SCANLON G. (eds.) *Fustat Expedition Final Report Vol. 2: Fustat-C*, Cairo (American Research Centre in Egypt).
84 BARNES 1997; and see also PFISTER R., 1938, *Les Toiles Imprimées de Fostat et de l'Hindoustan*, Paris; GITTINGER 1982; and IRWIN and HALL 1971.
85 BARNES 1997.
86 HURRY 1930.
87 See SCHREBER 1752.
88 Ibid.: 47–50.
89 British Library, Landsdowne MS 49, No. 58, f. 139.

90 MUNRO John H., 1983, 'The Medieval Scarlet and the Economics of Sartorial Splendour', in HARTE N.B. and PONTING K.G. (eds), *Cloth and Clothing in Medieval Europe*, London: 51ff.
91 WALTON P. 1991, 'Textiles', in BLAIR J. and RAMSAY N. (eds.), *English Medieval Industries*, London: 335.
92 CARDON Dominique, 1991, 'Black dyes for wool in Mediterranean textile centres: an example of the chemical relevance of guild regulations', *DHA* 9: 7–9.
93 Michael Bischof of Germany has produced many colours from different species of Turkish woad of various species dyed in different ways – see CARDON *et al.*, in press.
94 Lines from Chaucer's 'The Former Age', *c.*1380.
95 See conference papers published in BENNECKENSTEIN (ed.) 1988/89/91; MÜLLEROTT (ed.) 1990/92/93; and CARDON *et al.*, in press.
96 GUILLERE C., 'Trade and production of the pastel in Catalogne', and IGUAL LUIS David, 'The market of pastel in medieval Valence', both in CARDON *et al.*, in press.
97 CASADO ALONSO Hilario, 1990, 'El comercio del pastel. Datos para una geografia de la industria pañera española en el siglo XVI', *Revista de Historia Economica*, VIII, No. 3: 523–48.
98 ANON 1705: 306, 321.
99 See HURRY 1930: 94–116 for details.
100 Ibid.: 100.
101 ANON 1705: 312.
102 CASTER 1962: 138–57; BRUMONT Francis, 1994; 'La commercialisation du pastel toulousain (1350–1600)', *Annales du Midi*, 106, No. 205: 25–40; and WOLFF P., 1954, *Commerces et Marchands de Toulouse (vers 1350–vers 1450)*, Paris.
103 MICHEL Francisque, 1867, *Histoire de Commerce et de la Navigation à Bordeaux*, Bordeaux, Vol. 1: 295–310; HURRY 1930: 178–9; and CASADO ALONSO 1990 (n. 97): 523–48.
104 CROLACHIUM 1555; WEDELIUS 1675: Chs VII and VIII; SCHREBER 1752; and MÄGDEFRAU 1973. See also MÜLLEROTT 1992: Bibliography.
105 MÜLLEROTT 1992.
106 WATT 1890: 392.
107 HURRY 1930: 120–42.
108 BORLANDI Franko, 1959. 'Il commercio del guado nel Medioeveo, su 'Storia della Economia Italiana'', in *Saggi di Storia economica*, Torino. And see articles by BRUNETTI Antonello, BISCHI Delio and PETRONGARI Antonio in CARDON *et al.*, in press. Also BISCHI and PALOMBARINI Augusta in CARDON *et al.*, in press.
109 HURRY 1930: 158–60.
110 Ibid.: 276–83; and HARRIS Rendel, 1927, *A Primitive Dye-stuff* (Woodbrooke Essays, No. 10), Cambridge.
111 SALZMAN 1923: 203–5.
112 Ibid.: 209; PLOWRIGHT 1903: 97–8; and KOWALESKI Maryanne, 1993, *The Local Customs Accounts of the Port of Exeter 1266–1321*, Exeter (Devon and Cornwall Record Society, New Series, Vol. 36): 25–9.
113 HURRY 1930: 64–5; and PONTING 1976: 76.
114 See, for example, the Port Books of Southampton, Record Series, Reading University Library; GRAS N.S.B., 1918, *The Early English Customs System*, Oxford; CARUS-WILSON E.M. 1953, 'La guède française en Angleterre: un grand commerce du moyen age', *Revue du nord*, 35: 89–105; and HALL Hubert, 1885, *A History of the Custom-Revenue in England*, London, Vol. 1: 106, 280, 304.
115 KEENE Derek, 1985, *Survey of Medieval Winchester*, Oxford, Vol. 1: 303–4.
116 Ibid.: 315–18. For Scotland see GRIERSON 1986: 206–7; and for Coventry see SALZMAN 1923: 210–11.
117 BICKLEY Francis B. (ed.), 1900, *The Little Red Book of Bristol*, Bristol and London, Vol. 1: 54–5, and Vol. 2: 6–7, 16–22, 38–40, 170–74.
118 Ibid.: Vol. 2, 85–6.
119 See THIRSK Joan, 1978, *Economic Policy and Projects*, Oxford, for background information; and THIRSK 1997: 79–96.
120 HAKLUYT 1965, Vol. 2: 637.
121 MOTA Valdemar, 1991 (2nd edn), *o Pastel na Cultura e no Comércio dos Açores*, Ponta Delgada; and BENTLEY Duncan T., 1972, *Atlantic Islands, Madeira, the Azores and the Cape Verdes in Seventeenth Century Commerce and Navigation*, Chicago: 85–93. At the peak of production (1580–1630) the craze to grow woad even threatened food supplies, as it did in Britain. See also HAKLUYT 1965, Vol. 1: 148.
122 RHIND 1877: 508, notes the cultivation of woad here in the second half of the nineteenth century.
123 HAKLUYT 1965, Vol. 2: 752.
124 THIRSK 1997: 79–96; BETTEY 1978; CLARK and WAILES 1935–6; WILLS 1970; EDMONDS 1992: Appendix 1; and GRIERSON 1986: 206.
125 WILLS 1970. Displays and archives in Wisbech and Fenland Museum contain material relating to woad production in the area in the nineteenth and twentieth centuries.
126 For example, Hampshire Record Office, J.L. Jervoise, Herriard Collection: 44M66L/25/11–12 and 44M69/30/76. (I am grateful to staff at the HRO for their help.) See also BETTEY 1978: 115–17.
12 Interesting papers of 1800 in the Banks Stanhope collection, held at the Spalding Gentlemen's Society, relate to the formation of a woad company by a large group of Yorkshire dyers concerned at the inflated prices being asked by the producers of woad, who had become a cartel. I am grateful to the curator, J.W. Belsham, for drawing my attention to these. See also THIRSK 1997: 91–5.
128 WILLS 1970: 10, 16, 18–20.
129 Ibid.: 4, 14.
130 RHIND 1877: 508, and see *Sublime Indigo* 1987: 59–60.
131 CROOKES 1874: 453.
132 HURRY 1930: 289–94.

Chapter Three

Opening quote: A declaration by E. de-Latour of the Bengal Civil Service and magistrate of Faridpur in 1848 (reported in 'Papers Relating to Indigo Cultivation in Bengal', National Archives of India).

1 CHAUDHURI K.N., 1985, *Trade and Civilisation in the Indian Ocean: An Economic History from the Rise of Islam to 1750*, Cambridge: 90.
2 CHAUDHURI 1965: 8.
3 CHAUDHURI 1985 (n. 1): 82.
4 BOYAJIAN James C., 1993, *Portuguese Trade in Asia under the Habsburgs, 1580–1640*, Baltimore and London.
5 Ibid.: 46.
6 BECKMAN 1846: 274.
7 WATT 1890: 391–2; for some Dutch sources see ALDEN 1965: 38–9.
8 See, for general background, FOSTER William, 1933, *England's Quest of Eastern Trade*, London.
9 PIRÉS Tomé, 1944, *The Suma Oriental of Tomé Pirés* (trans. and ed. CORTESÃO Armando), London, Vol. 1: 43.
10 BARBOSA Duarte, 1918, *The Book of Duarte Barbosa*, (trans. LONGWORTH DAMES Mansel), London (Hakluyt Soc., 2nd Series, 44), Vol. 1: 154.
11 HAKLUYT 1965, Vol. 1: 210. For general background see JANAKI V.A., 1980, *The Commerce of Cambay from the Earliest Period to the Nineteenth Century*, Baroda.
12 CHAUDHURI 1965: 176; BIRDWOOD George (ed.), 1843 (reprint 1965), *The Register of Letters of the Governor and Company of Merchants of London trading into the East Indies, 1600–1619*, London: 255, 287, 337, 405; and GOPAL Surendra, 1975, *Commerce and Crafts in Gujarat in the 16th and 17th Centuries*, New Delhi.
13 GITTINGER 1982: 22.
14 FOSTER 1968: 152–4. And see nn. 19 and 20, below.
15 FOSTER 1968: 174, 207; TAVERNIER Jean-Baptiste, 1889, *Travels in India*, London, Vol. 1: 69 and Vol. 2: 9; and see 'Anile, neel' in the glossary compiled by YULE Henry and BURNELL A.C. (eds), 1903 (reprint 1986), *Hobson Jobson*, London.
16 PELSAERT 1925: 30; and see GOPAL 1975 (n. 12): 197–8.
17 TRIVEDI K.K., 1994, 'Innovation and Change in Indigo Production in Bayana, Eastern Rajasthan', *Studies in History*, 10, 1: 69.
18 *JSCI* 1902, Vol. XXI: 1204.
19 PELSAERT 1925: 10–18; and see TRIVEDI K.K., 'Comparative Systems of Indigo Production in Pre-British India', in CARDON *et al.*, in press.
20 TAVERNIER 1889 (n. 15), Vol. 2: 10–12.
21 DAS GUPTA 1979: 58.
22 PELSAERT 1925: 16.
23 FOSTER 1968: 123–5, 148–55.
24 Ibid.: 174.
25 CHAUDHURI 1985 (n. 1): 200; see also PELSAERT 1925: 16–17; FAWCETT 1836: 226, 258; and GOPAL 1975 (n. 12): 223–5, 235–6.
26 TAVERNIER 1889 (n. 15) Vol. 2: 31; SAINSBURY 1916: 262; and DAS GUPTA 1979: 60.
27 SAINSBURY 1909: 108–9; and HARLEY 1982: 201–2.
28 IRWIN John and SCHWARTZ P.R., 1966, *Studies in Indo-European Textile History*, Ahmedabad: 13–15; GOKHALE Balkrishna Govind, 1979, *Surat in the Seventeenth Century*, London; and FAWCETT 1936.
29 CHAUDHURI 1965: 176–88; CHAUDHURI 1978: 330–35, 523 (Appendix 5); SAINSBURY 1907–22; and FOSTER and FAWCETT, 1906–55 (particularly FAWCETT 1936). The original EIC Court minutes are in London – see the Bibliography (unpublished works) for details.
30 SAINSBURY 1912: e.g. v, 107, 149, 186; and SAINSBURY 1916: e.g. 117–18.
31 WATT 1890: 392.
32 See SAINSBURY 1916 and 1922 for refs.
33 BRUCE J., 1810, *Annals of the Honorable East India Company from their establishment by the charter of Queen Elizabeth, 1600, to the Union of the London and English East India Companies, 1707–8*, London, Vol. 3: 182.
34 CHAUDHURI 1978: 523 (Appendix 5).
35 GITTINGER 1982.
36 IRWIN and BRETT 1970: 1 and see 3–6.
37 CHAUDHURI 1985 (n. 1): 82.
38 TAVERNIER 1889 (n. 15), Vol. 1: 9; and see CHAUDHURI Susil, 1975, *Trade and Commercial Organization in Bengal, 1650–1720*, Calcutta.
39 Diary No. 109, 24 April 1801, Secret and Political Dept. of the Bombay Government, Bombay Record Office.
40 See BALFOUR-PAUL 1997: 37–41 for refs; and RHIND 1877: 505.
41 BRAUDEL Fernand, 1972, *The Mediterranean and the Mediterranean World in the Age of Philip II*, London, Vol. 1: 567.
42 See BALFOUR-PAUL 1997: 30–34 for refs.
43 BERCHET G., 1896, *Le Relazioni dei Consoli Veneti nella Siria*, Turin: 79–80.
44 CHAUDHURI 1965: 11.
45 FOSTER William (ed.), 1931, *The Travels of John Sanderson in the Levant, 1584–1602*, London: 189.
46 See ROBERTS L., 1974, *The Merchants Mappe of Commerce*, Amsterdam (facsimile of the original, 1638, in Cambridge University library): 118, 136, 221.
47 IRWIN and SCHWARTZ 1966 (n. 28): 8–13.
48 See BALFOUR-PAUL 1997: 36 for refs.
49 RAMBERT Gaston, 1966, *Histoire du Commerce de Marseille, 1660–1789*, Paris, Vol. 7: 173, and see Vol. 6 (1959): 403–6.
50 VOLNEY C.F., 1897, *Travels through Syria and Egypt*, Vol. 2: 406–31.
51 FAROQUI Suraiya, 1984, *Towns and townsmen of Ottoman Anatolia*, Cambridge: 146.
52 PARKINSON 1640: 601.
53 SAHAGÚN Bernardino de, 1985, *Historia General de las Cosas de Nueva España*, (ed. Porrúa), Mexico: 699; and HAKLUYT 1965, Vol. 2: 587.
54 HAKLUYT 1965, Vol. 2: 590, 593.
55 Ibid.: Vol. 2: 454 (and for the following quotes).
56 RHIND 1877: 494–7; CARDON and du CHATENET 1990: 114–15; and KNECHT, RAWSON and LOWENTHAL 1941, Vol. 1: 330–49.

57 HURRY 1930: 267–8; and WATT 1890: 392.
58 VETTERLI 1951: 3068; and RHIND 1877: 500.
59 HURRY 1930: 268–71.
60 Ibid.: 270; RHIND 1877: 500; and WATT 1890: 393.
61 PARKINSON 1640: 600–601.
62 RHIND 1877: 501.
63 CARDON Dominique in *Rouge, Bleu, Blanc. Teintures à Nîmes* 1989: 72.
64 GIOBERT 1813: 219–65; and PUYMAURIN 1810.
65 See BALFOUR-PAUL 1997: 49–52 for details.
66 DE LANESSAN 1886: 873, 877–8.
67 Lousiana council minutes quoted in HOLMES 1967: 331.
68 Ibid.
69 MONNEREAU 1769: 60–61.
70 GRAY 1958, Vol. I: 120.
71 ARONSON Lisa, 1980, 'History of Cloth Trade in the Niger Delta: A Study of Diffusion', *Textile History*, Vol. 11: 89–107. And see BALFOUR-PAUL 1997: 18, 20, 40.
72 THOMAS 1997: 317–20.
73 Ibid.: 62, 334; and BOSER-SARIVAXÉVANIS 1969: 11, 154, 321.
74 THOMAS 1997: 139–44.
75 ARONSON 1980 (n. 71): 92.
76 MOLLIEN Gaspar Theodore, 1967 (reprint), *Travels in the Interior of Africa to the Sources of the Senegal and Gambia…*, London: 256.
77 PARK Mungo, 1815, *The Journal of a Mission to the Interior of Africa in the Year 1805*, London: 11 and Appendix VI.
78 DE LANESSAN 1886: 122.
79 JOHNSON Marion, 1980, 'Cloth as Money: the Cloth Strip Currencies of Africa', *Textile History*, Vol. 11: 198–9.
80 EQUIANO Olaudah, 1995, *The Interesting Narrative and Other Writings* (ed. CARRETTA Vincent), London: 235.
81 DAVIES 1957, *The Royal African Company*, London: 220–21.
82 RHIND 1877: 500–501.
83 HARDY Georges, 1921, *La mise en valeur du Sénégal de 1817 à 1854*, Paris: 160, 164–6; de LANESSAN 1886: 118–20.
84 CAILLIÉ René, 1830, *Journal d'un Voyage à Temboctou et à Jenné*, Paris, Vol. 1: 32, 45, 106.
85 Surviving American Indian methods of indigo dyeing have been studied by Ana Roquero, dye researcher in Madrid.
86 MACLEOD 1973: 177–8.
87 ARNOLD 1987 and REMBERT 1979.
88 MACLEOD 1973: 176–203, 222; SMITH 1959. See also MOZIÑO 1826; RUBIO 1976; and FERNANDEZ MOLINA 1992 (unpublished).
89 OSBORNE Lilly de Jongh, 1965, *Indian Crafts of Guatemala and El Salvador*, Oklahoma: 36–7; MACLEOD 1973: 180–81; WORTMAN Miles L., 1982, *Government and Society in Central America 1680–1840*, New York: esp. 178; and LUTZ Christopher H., 'The Late Nineteenth-Century Guatemalan Maya in Historical Context: Past and Future Research', in SCHEVILL Margot Blum, 1993, *Maya Textiles of Guatemala*, Texas: Ch. 4.
90 SMITH 1959: 200.
91 Pers. comm., Ivor Noël Hume of Williamsburg.
92 HOLMES 1967: 334.
93 Ibid.: 193–208.
94 RAYNAL Abbé (trans. JUSTAMOND M.A.), 1777, *A Philosophical and Political History of the Settlements and Trade of the Europeans in the East and West Indies*, London (3rd edn), Vol. II: 417.
95 MACLEOD 1973: 184–93; and SMITH 1959: 183–96.
96 SMITH 1959: 181, 210–11.
97 BISCHOF Michael, 1996, 'Natural Indigo from El Salvador', *Textileforum* 2: 26–7, and pers. comm. over several years.
98 PINO Raúl del, 1895, 'Del cultivo y elaboración del añil en al Estado de Chiapas', *La Tierra*; and DIAZ Gerardo Sánchez, 1991, 'Cultivo, producción y mercado del añil en México en Michoacán en el siglo XIX', *Nuestra Historia*, Revista historiográfica, No. 1 (Caracas): 38–40.
99 TUROK Martha, 1996, 'Textiles de Oaxaca', *Artes de Mexico*, No. 35: 67, 93; and pers. comm. with the author.
100 See ALDEN 1965: 35–60 for the following information.
101 Ibid.: 60.
102 DU TERTRE, Jean-Baptiste, 1658, *L'Histoire des Antilles*, 1658; and see RHIND 1877: 500.
103 THOMAS 1997: 257; SLOANE 1725: 34–7; and GRAY 1958: 54, 291.
104 MILBURN 1813, Vol. II: 213–14.
105 DE LANESSAN 1886: 428, 493
106 VETTERLI 1951: 3069.
107 MICHEL Francisque, 1870, *Histoire du Commerce et de la Navigation à Bordeaux*, Vol. 2: 286–8; and RAMBERT 1959: 403–6.
108 DE BEAUVAIS-RASEAU 1770: Plate VII.
109 BISHOP 1868, Vol. 2: 314, 348.
110 GRAY 1958
111 Ibid.: 73–4; BISHOP 1868: 348; and see HOLMES 1967.
112 HOLMES 1967: 336–7.
113 PETTIT 1974: 43–4.
114 HOLMES 1967: 341–3; GRAY 1958: 293–7, 1024; and BISHOP 1868: 349.
115 COON 1976; SHARRER 1971; and BISHOP 1868: 348–9.
116 GRAY 1958: 1024; and. ROGERS 1970: 90–91.
117 ROGERS 1970: 88 and see 52–3, 85–92.
118 Quoted in THOMAS 1997: 268.
119 THOMAS 1997: 533; and DAVIES 1957 (n. 81): 187, 340, 360 (Appendix II).
120 E.g REID 1887; CRAWFURD John, 1831, *Letters from British Settlers in the Interior of India*, London; and BEAMES John, 1961, *Memoirs of a Bengal Civilian*, London.
121 RAWSON 1899: 174.
122 NIGHTINGALE Pamela, 1970, *Trade and Empire in Western India, 1784–1806*, Cambridge: 136, 138, 195.
123 WATT 1890: 393–6, and KLING 1966: 17–19. Also SINHA 1965/70: Vol. 1: 97–100, 206–11, 227–30; and Vol. 3: 19–22.
124 East India Company 1836 (Indigo): 32.
125 RAWSON 1899: 173–4; and see East India Company 1836.
126 TRIPATHI Amales, 1979, *Trade and Finance in the Bengal Presidency, 1793–1833*, Calcutta: 183–5, 217–19.
127 KLING 1966: 29–37; and see HEBER Reginald, 1828, *Narrative of a journey through the Upper Provinces of India from Calcutta to Bombay, 1824–5 etc.*, London, Vol. 1: 113.

128 KLING 1966: 144–5.
129 *Reminiscences of Behar*, 1887/88: 39–43.
130 KLING 1966: 52–55.
131 KAYE J.W., text accompanying William Simpson's illustration of a Bengali indigo factory, in *India, Ancient and Modern*, London 1867, Vol. 2: 76–8.
132 Ibid.: 190–92; and KLING 1966: 41–4, 104.
133 East India Company 1836 (Indigo): 37; MILBURN 1813, Vol. 1: 288–91; and KLING 1966: 20–22.
134 RHIND 1877: 504.
135 BECKMAN 1846: 280.
136 BHATTACHARYA Subhas, 1977–8, 'Rent Disturbances of 1860–62 and the Indigo Revolt', *Quarterly Review of Historical Studies*, 17 (4): 214–20; see also *Reminiscences of Behar* 1887/88: 39–43.
137 KLING 1966; MOLLA 1973; and CHAKLADAR 1960, 1972 and 1973.
138 TENDULKAR 1957: 5.
139 WATT 1890: 379; and KLING 1966: 124–46.
140 KLING 1966: 121, and see 111–24
141 For a biography see ODDIE Geoffrey, 1998, *James Long of Bengal – Missionary, Scholar and People's Hero*, London.
142 RAO Amiya, 1991, '*Neel Darpan – The Story of Indigo*', *IASSI Quarterly*, Vol. 10, No. 1.
143 BASU 1903: Appendix B.
144 LONG James, 1861, *Strike, but Hear! Evidence Explanatory of the Indigo System in Lower Bengal*, Calcutta.
145 SIMKIN C.G.F., 1968, *The Traditional Trade of Asia*, London: 288, 294.
146 *Brahmins and Pariahs*, 1861, Calcutta.
147 See WATT 1890: 400–416, for the different places.
148 MITTAL 1978.
149 RAWSON 1899; see also SINHA B.K., 1976: 641–8.
150 WATT 1890: 396, 398–9, 413; and DAVIS 1918, Part 1: 38–9.
151 BURKILL 1935: 1233.
152 RAWSON 1899: 473; DAVIS 1918, Part 1: 32; and see *JSCI*, 1902, Vol. XXI, May: 731.
153 DE LANESSAN 1886: 121–2.
154 Pers. comm., 1980s, Cuthbert Skilbeck, retired drysalter and honorary archivist to the Dyers' Company of London.
155 *Wealth of India*, 1959: 173.
156 PERSOZ 1846, Vol. 1: 421–58.
157 ALDEN 1965: 43
158 RAFFLES 1994, Vol. 1: 212.
159 Ibid., Vol. 2: 256; and see CRAWFURD 1820, Vol. 3: 356, and Vol. 1: 460.
160 CRAWFURD 1820 (n. 159), Vol. 3: 356, and Vol. 1: 460.
161 Ibid.: Vol. 1: 460–61; BURKILL 1935: 1234; and *JSCI*, 1907, April: 390, 457.
162 HEMKER Marijke, 1985, 'Het Blauwe zweet van de Javaan', in *Indigo, Leven in een kleur* (ed. OEI Loan), Weesp: 121–2.
163 *JSCI*, 1899, Vol. XVIII: 798; and DAVIS 1918, Part 1: 39–40.
164 RAFFLES 1994, Vol. 1: 205.
165 KLING 1966: 116–21, 219–23; and MITTAL 1978.
166 BEAMES 1961 (n. 20): 170–74, 182–4; and MITTAL 1978: 66–70.
167 TENDULKAR 1957; and MITTAL 1978.
168 SMITH 1959: 210.
169 ROBERTS Richard, 1984, 'Women's Work and Women's Property: Household Social Relations in the Maraka Textile Industry of the Nineteenth Century', *Comparative Studies in Society and History*, Vol. 26, No. 2: 229–50.
170 BALFOUR-PAUL 1997: 53–61.
171 SPRING Christopher and HUDSON Julie 1995, *North African Textiles*, London: 99–104; and PICTON and MACK 1969: 169.
172 Presidential address to the Society of Chemical Industry – see *JSCI*, 1898, Vol. XVIII: 474.
173 RAWSON 1899: 174.
174 MCLAREN 1986: 11–19.
175 Information from Zeneca. And see indigo patents in the *JSCI*, e.g. 1902, Vol. XXI: 38–9, 1072–3.
176 Davis 1918, Part 1: 33–7; SEEFELDER 1994: 43–68; and SCHMIDT Helmut, 1993, 'Indigo – 100 years of industrial synthesis', *Chemie*, No. 3: 121–8.
177 *JSCI*, Vol. XVIII: 1068.
178 REED 1992: 124.
179 See the trade reports in the *JSCI* and *JSDC* from 1898, e.g. *JSCI*, 1902, Vol. XXI: 1111, 1045.
180 Information from Zeneca; and see DAVIS 1918, Part 1: 37; and REED 1992: 116–23.
181 Pers. comm., Graham Ashton, BASF denim team (Zeneca); and Zeneca literature.
182 *JSCI*, 1902, Vol. XXI, May: 648.
183 Ibid.: 80; and *JSCI*, 1899, Vol. XVIII: 718.
184 *JSCI*, 1902, Vol. XXI: 1111.
185 *JSDC*, 1898, Vol. XIV: 122; see also 15–16, 26–7, 36, 136, 180; and Vol. XV, 1899: 50.
186 MARTIN-LEAKE 1975: 367.
187 *Indigo Publications* 1918–23; and *Indigo Research Reports* 1906–13.
188 DAVIS 1918, Part 1: 39.
189 DAVIS 1918, Part 1: 33, 40–46; GHOSH 1944 (No. 11): 491; and BURKILL 1935: 1236, 1239.
190 Ibid., Part 3: 441–59; and see GHOSH 1944 (No. 12): 537–42.
191 *JSCI*, 1899, Vol. XVIII: 267–8.
192 MCLAREN 1986: 18; and DAVIS 1918: 43–5.
193 ICI literature on 'Indigo LL on Cotton'.

Chapter Four

Opening quote: Crawfurd John, 1820, *History of the Indian Archipelago*, Edinburgh, Vol. 1: 461.

1 See BALFOUR-PAUL 1997: 45–6 for refs.
2 Pers. comm., botanists Brian Schrire and Martyn Rix. See also CARDON and DU CHATENET 1990: 140–59, 212–19, 257–8; *Sublime Indigo* 1987: 38–58; and MOOK-ANDREAE 1985: 13–20.
3 GILLET 1958 and SCHRIRE 1995.
4 *Sublime Indigo* 1987: 43–5.
5 CHRISTIE, in press.
6 LOMBARD Maurice, 1978, *Les Textiles dans le Monde Musulman*, Paris: 134–43.
7 BURKILL 1935: 1234; and CRAWFURD John, 1820, *History of the Indian Archipelago*, Edinburgh, Vol. 1: 458.

8 BALFOUR-PAUL 1997: 42–67.
9 KRÜNITZ 1783: 558; and DE BEAUVAIS-RASEAU 1770: 8–9.
10 HEYD W., 1923 (reprint), *Histoire de Commerce du Levant au Moyen-Age*, Leipzig, Vol. 2: 10, 628; and RHIND 1877: 501.
11 Pers. corr. Brian Schrire, Royal Botanic Gardens, Kew.
12 BURKILL 1935: 1234, 1239; and LEMMENS and WULIJARNI-SOETJIPTO 1991: 82. The present writer found it being used in southern China in 1993.
13 SCHWEPPE 1992: 293.
14 ARNOLD 1987: 53–83 and REMBERT 1979: 128–34.
15 PARKINSON 1640: 601.
16 HURRY 1930: 1–10.
17 GERARD John, 1597, *The Herball*, London: 394.
18 HILL 1992: 23 and Tables 1–2.
19 RHIND 1877: 508.
20 BISCHOF Michael, 'Pastel-Isatis-Civit Otu in Turkey', in CARDON *et al.*, in press.
21 NEEDHAM Joseph, 1986, in NEEDHAM Joseph (ed.), *Science and Civilisation in China* Cambridge, Vol. 6, Part I: 158.
22 FORTUNE Robert, 1846, 'A Notice of the Tien-ching, or Chinese Indigo', *Royal Horticultural Society Journal*: 269–71; and his *Two Visits to the Tea Countries of China*, London 1853, Vol. 1: 184, 234–5.
23 YING-HSING 1966: 76.
24 WATT 1890: 524.
25 NEEDHAM 1986 (n. 21): 158.
26 LEMMENS and WULIJARNI-SOETJIPTO 1991: 137; and REIN 1889: 173.
27 JOLY 1839; BÉRARD 1838; REIN 1889: 175; and SCHUNK 1855, Part 1: 76–7.
28 HILL 1992: 24–6.
29 SPOONER 1943: 102–17; and see BALFOUR-PAUL 1996: 94–9.
30 LEMMENS and WULIJARNI-SOETJIPTO 1991: 138; WATT 1890: 376–7; MOHANTY *et al.*, 1987: 153–4; and NAKAO Sasuke and NISHIOKA Dasho Keiji, 1984, *Flowers of Bhutan*, Tokyo: 138–9.
31 WATT 1890: 416.
32 XIA Zhi-Qiang and ZENK Meinhart H., 1992, 'Biosynthesis of Indigo Precursors in Higher Plants', *Phytochemistry*, Vol. 31, No. 8: 2695.
33 Possibly due to co-pigmentation – research is currently being undertaken by Professor Claude Andary at Montpellier University.
34 LEMMENS and WULIJARNI-SOETJIPTO 1991: 135.
35 CAILLIÉ, René, 1830, *Journal d'un Voyage à Temboctou et à Jenné*, Paris, Vol. 2: 57–8.
36 MARSDEN 1811: 94.
37 LEMMENS and WULIJARNI-SOETJIPTO 1991: 93; and CARDON and DU CHATENET 1990: 155, 214–15.
38 WATT 1890: 396; East India Company 1836: 91; and MOHANTY *et al.*, 1987: 34–5.
39 ROXBURGH 1811.
40 WATT 1890: 451.
41 BRANDON 1986: 43–4.
42 BOND 1996 (unpublished).
43 E.g. KRÜNITZ 1783: 569.
44 CARDON and DU CHATENET 1990: 157–9, 257–8.
45 *Sublime Indigo* 1987: 41.
46 BRAY Francesca, 1984, 'Agriculture' in *Science and Civilisation in China* (ed. NEEDHAM Joseph), Cambridge, Vol. 6, Part II: 277, 279, 509, 593.
47 BALFOUR-PAUL 1996: 95.
48 MOHANTY *et al.*, 1987: 83.
49 The Centre for Economic Botany at the Royal Botanic Gardens, Kew, London, has a large collection of indigo material from all over the world.
50 WATT 1890: 399–427.
51 RAFFLES 1994, Vol. 1: 132.
52 *JSCI*, Vol. XXI: 1111; HOWARD and HOWARD 1914; and DAVIS 1918, Part 2: 207–13.
53 MARTIN-LEAKE 1975: 365.
54 CRAWFURD 1820 (n. 7) Vol. 1: 457–61.
55 WATT 1890: 399–416; and RAWSON 1899: 467.
56 MARTIN-LEAKE 1975: 362–3.
57 FOSTER 1968: 154.
58 PELSAERT 1925: 12.
59 MOEYES 1993: 41.
60 SPOONER Roy C. *et al.*, 1943, 'Indican content of Szechwan indigo and the effect of fertilisers', *Journal of Chinese Chemical Society*, 10: 69–76.
61 HILL 1992: 24–6 and Tables 4–8. See also REIN 1889: 174; and DAVIS 1918, Part 3: 453–6.
62 HURRY 1930: 16–17, 288.
63 MARTIN-LEAKE 1975: 370; MOHANTY *et al.*, 1987: 83; and *Wealth of India* 1959, Vol. V: 183.
64 WATT 1890: 401.
65 See, for example, *Reminiscences of Behar* 1887/88: 117.
66 WILLS 1970.
67 Susanna Peckover manuscript, 'Woad, its History and Cultivation', in the archives of Wisbech and Fenland Museum, Cambridge. Other relics of England's woad industry exist in Lincolnshire (Museums of Boston and the Gentlemen's Society, Spalding) and London (Science Museum).
68 BALFOUR-PAUL 1997: 50.
69 *Scientific American Supplement*, 1888, No. 667: 10649.
70 ERB Maribeth, 'The Curse of the Cooked People', in HAMILTON 1994: 206.
71 DIDEROT 1765, Vol. 8: 683.
72 LEGGETT 1944: 21.
73 RAYNAL Abbé (trans. JUSTAMOND M.A.), 1777, *A Philosophical and Political History of the Settlements and Trade of the Europeans in the East and West Indies*, London (3rd edn), Vol. II: 417–18.
74 SCHUNK 1855–7: 73–4.
75 Ibid.: 73–95. And see SCHWEPPE 1992: 286.
76 NAPIER 1875: 232–5.
77 For Molisch's experiments see *JSCI*, 1899, Vol. XVIII: 361; see also RAWSON 1899: 171–2.
78 Ibid.: 471.
79 Research undertaken by Kerry Stoker, David Hill and David Cooke at Long Ashton Research Station, University of Bristol.
80 XIA and ZENK 1992 (n. 32): 2695–7; and MAIER Walter, SCHUMANN Brigitte and GRÖGER Detlef, 1990, 'Biosynthesis of Indoxyl derivatives in *Isatis tinctoria* and *Polygonum tinctorium*', *Phytochemistry*, Vol. 29, No. 3: 817–19.

81 PLOWRIGHT 1903: 108–10.
82 See n. 79.
83 BÜHLER 1951: 3088–91; and present writer's fieldwork.
84 These are at the French Château-Musée de Magrin, east of Toulouse, and at Molschleben and Gotha in Thuringia, Germany. See MÜLLEROTT 1992; and BALFOUR-PAUL 1993: 4–5.
85 WILLS 1970: 16–20; also his 'Woad Growing and Manufacture in the Lincolnshire and Cambridgeshire Fens and in Northhamptonshire', *DHA* 11, 1993: 6–9.
86 HURRY 1930: 22–31.
87 WILLS 1970: 15.
88 There are photographs, paintings and models in the British museums mentioned in n. 67. See also HURRY 1930: plates XI–XIII; SCHREBER 1752; and WILLS 1970.
89 SANDBERG 1989: 26–9.
90 HURRY 1930: 30–31, 135.
91 ANON 1905: 439.
92 BICKLEY Francis B. (ed.), 1900, *The Little Red Book of Bristol*, Bristol and London, Vol. 1: 54–5.
93 'Rammers' were also used to pack the couched woad into casks – there is one in the Science Museum, London.
94 BETHE 1884: 73–4. For smaller-scale processing see WADA *et al.*, 1983: 277.
95 AMANO Masatoshi, 'Development of Awa Indigo Industry in the Latter Half of Tokugawa Period and Economic Modernization of Tokushima Prefecture', in MÜLLEROTT (ed.) 1992: 30–41.
96 BRANDON 1986: 47.
97 UEDA Toshi, 1975, *History of Ai in Tokushima* (in Japanese), Tokushima. See YONEKAWA Takahiro, 'The Japanese Natural Indigo Dye; its Present and Future Challenges, in MÜLLEROTT (ed.) 1993: 18–21; and SCHUNK Edward, 'Etudes microscopiques sur le gisement de la matière bleue dans les feuilles du Polygonum tinctorium', *Mêmoire lu à l'Académie des Sciences, 12 Nov. 1838*. There is a splendid display of local indigo manufacture in Aizumicho Historical Museum in Tokushima.
98 WEDELIUS 1675: 40.
99 EWERDWALBESLOH Imke and MEYER Ortwin, 'Bacteriology and Enzymology of the Woad Fermentation', in CARDON *et al.*, in press. Research has also been undertaken by Dr Satoshi Ushida at Mukogawa Women's University and by Professor Phillip Johns at Reading University.
100 BARBOT Jean, 1752, 'A Description of the Coasts of North and South Guinea…', in CHURCHILL A. *et al.*, *A Collection of Voyages and Travels*, Vol. V: 41; and DENHAM D. and CLAPPERTON H., 1826, *Narrative of Travels and Discoveries in Central Africa in the Years 1822, 1823 and 1824*, London: 60, 174; and CLAPPERTON H., 1829, *Journal of a Second Expedition into the Interior of Africa from the Bight of Benin to Soccatoo*, London: 173–4.
101 NAKAO and NISHIOKA 1984, (n. 30): 138–9.
102 SEWELL W.G. *et al.*, 1939, 'The Natural Dyes of Szechwan, West China', *JSDC*, Vol. 54: 415; and SPOONER 1943: 112.
103 CRAWFURD 1820 (n. 7), Vol. 1: 461.
104 CROOKES 1874: 451, 456.
105 STUART G.A., 1979, *Chinese Materia Medica*, Taipei (original edn published Shanghai 1911): 218.
106 JSCI, 1899, Vol. XVIII: 534.
107 *JSCI*, 1902, Vol. XXI: 1420.
108 *Scientific American Supplement*, 1888, No. 667: 10649; WILSON Ernest Henry, 1913, *A Naturalist in Western China*, London, Vol. 2: 85; YING-HSING 1966: 76; and SEWELL *et al.*, 1939 (n. 102): 414.
109 ROSSI Gail, 1988, 'Growing indigo in China's Wanchao district', *Surface Design Journal*, Summer 1988: 28–9; ROSSI 1990: 20–23; and BALFOUR-PAUL 1996: 95.
110 MARSDEN 1811: 94; and MOEYES 1993: 54.
111 Information supplied by Kiichirou Ono of Toyama Memorial Museum.
112 RAFFLES 1994, Vol. 1: 132–3.
113 At Bharatpur, near Agra, there are still some remains of old indigo tanks.
114 PELSAERT 1925: 10–12; and FOSTER 1968: 151.
115 FOSTER 1968: 151.
116 Ibid.: 154.
117 For Indian systems see MOHANTY *et al.*, 1987: 27–36.
118 See, for example, introduction to the translation of CHARPENTIER-COSSIGNY DE PALMA 1789.
119 BALFOUR-PAUL 1997: 41, 53–5.
120 WATT 1890: 417–19; and MONNEREAU 1769: 7–9.
121 DIDEROT 1765: Vol. 8: 680.
122 MONNEREAU 1769: 12 and see 17–41.
123 DIDEROT 1765, Vol. 8: 681, and 'Indigoterie' Plate in Vol. 22.
124 CROOKES 1874: 448; and SLOANE 1925: 35.
125 *Reminiscences of Behar* 1887/88: 35.
126 MACLEOD 1973: 180.
127 MARTIN-LEAKE 1975: 366.
128 DIDEROT 1765, Vol. 8: 680–81.
129 CHARPENTIER-COSSIGNY DE PALMA 1789: 8 and see 9–85.
130 RAWSON 1899: 467–73; DAVIS 1918, Part 2: 213–21; CROOKES 1874: 448–50; and see MOHANTY *et al.*, 1987: 83–90.
131 HALLER 1951: 3074; and WATT 1890: 446, 448.
132 ANON 1705: 312–14.
133 E.g. East India Company 1836: 67, 70.
134 MILBURN 1813, Vol. 1: 290–91.
135 CROOKES 1874: 457.
136 NAPIER 1875: 235–482; PERSOZ 1846, Vol. 1: 426–58; and KNECHT RAWSON and LOWENTHAL 1941: 815–29.
137 The KÖK carpet project is the brainchild of Michael Bischoff (see his article in CARDON *et al.*, in press).
138 'Renaissance Dyeing' is owned and run by David and Margaret Redpath in Pembrokeshire.
139 *Koreana*, 1997, Vol. 11, No. 2: 66–71.
140 Return of Traditional Dyes in Guinea, IDRC Reports, April 1995.
141 Projects undertaken at Reading University by a team led by Philip John, and at Bristol University (see n. 79). See HILL 1992: 23–6; and STOKER 1997 (unpublished). Jill Goodwin, author of *A Dyer's Manual*, has long been a passionate advocate of woad.
142 E.g. PARKINSON 1640: 601; and CULPEPER 1826, *Complete Herbal and English Physician*, Manchester (facsimile, Harvey Sales 1981): 200.

Chapter Five

Opening quote: MAIRET 1916: 8.

1 For insolubilisation see MCLAREN 1986: 58.
2 HENDERSON P. 1950, *The Letters of William Morris to his Family and Friends*, London: 76.
3 BANCROFT 1813, Vol. 1: 225–41; CRACE-CALVERT 1876: 175–81; MAIRET 1916: 63–68.
4 BRÜGGEMANN and BÖHMER 1982: 93, 115.
5 BOND 1996 (unpublished).
6 SEEFELDER 1994: 34–37.
7 Pers. comm., Jurek Mencel, former indigo consultant to ICI, and Graham Ashton of BASF (Zeneca plc). See also BASF technical literature.
8 Chemically speaking, the indigo is reduced as the vegetable matter decomposes with the liberation of hydrogen. Saccharide-producing carbohydrates convert by fermentation to lactic acid, thence into butyric acid, carbon dioxide and hydrogen. Alkali such as slaked lime or lye provides the soluble salt of the reduced indigo. (Information from Zeneca.)
9 The late Nancy Stanfield lived in western Nigeria in the 1950s and 60s and studied local indigo dyeing there (see STANFIELD 1971). She generously shared information and provided photographs during memorable visits arranged by Jacqueline Herald.
10 HAMILTON 1994: 62.
11 *Arte della Lana*, Biblioteca Riccardiana, Florence, Codex 2580, f. 141v–147, February 1418.
12 CARDON Dominique, 'The medieval woad-vat: two previously unpublished sources', in CARDON *et al.*, in press; and CARDON 1992a: 22–31.
13 CARDON forthcoming.
14 CARDON 1992b: 12–13.
15 Ibid.: 10–11. And see REBORA 1970: 66–7.
16 REBORA 1970: 84.
17 BARBOT 1732: 41.
18 STANFIELD 1971; and POLAKOFF 1982: 25–6.
19 Writer's fieldwork in 1997, and see *Teinture, expression de la tradition Afrique noire* 1982: 10; and WAHLMAN and CHUTA 1979: 459.
20 PARK Mungo, 1815, *The Journal of a Mission to the Interior of Africa in the Year 1805*, London: 10–12. See also CAILLIÉ René, 1830, *Journal d'un Voyage à Temboctou et à Jenné*, Vol. 2: 57.
21 De NEGRI 1966: 97.
22 For Guinea, for example, see de LESTRANGE Monique, 1950, 'Les Sarankoles de Baydar (techniques de teinturiers)', *Études Guinéennes* (Conakry): 22–3.
23 WADA *et al.*, 1983: 277.
24 Writer's fieldwork. See also USHIDA Satoshi, 1992, 'Indigo dyeing in Japan', *DHA* 10: 32–4; TAKAHARA Yoshimasa, 'Biological Reduction of Indigo', CARDON *et al.*, in press; BETHE 1984: 74–5; and BRANDON 1986: 50–51.
25 Dr Ana Roquero of Madrid in Topochico, Imbabura.
26 For 'sweet' vats of India see MOHANTY *et al.*, 1987: 29–30.
27 CARDON and du CHATENET 1990: 257–8. *Sacatinta* is a type of 'false' indigo which produces a pale blue direct dye. See SAYER 1985: 144. In Guatemala in 1996 Cati Ramsay found the last indigo dyer near Quetzaltenango still using *sacatinta* (renewed every two months) in an indigo vat otherwise made from modern ingredients.
28 MOEYES 1993: 57–8.
29 ERB Maribeth, 'The Curse of the Crooked People' in HAMILTON 1994: 206.
30 See, for example, THEISON 1982: 72–4, for Sumatra; HERINGA 1989 for Java.
31 See, for example, DANIEL 1938; BYFIELD Judith, 1994, 'Technology and Change: The Incorporation of Synthetic Dye Techniques in Abeokuta, Southwestern Nigeria', *Contact, Crossover, Continuity: Proceedings of the Fourth Biennial Symposium of The Textile Society of America*: 45–51; and BYFIELD 1993 (unpublished).
32 Writer's fieldwork, and see PENLEY 1988: 140–43.
33 CANNIZZO Jeanne, 1983, 'Gara Cloth by Senesse Tarawallie', *African Arts*, Vol. 16, No. 4: 60.
34 BOSENCE 1985: 42, 48.
35 HALLEAUX 1981: 34–5.
36 LAGERCRANTZ Otto, 1913, *Papyrus Holmiensis*, Upsala: 211–12.
37 Lichen dye was used in antiquity to imitate shellfish purple, and in England until the nineteenth century was combined with indigo to economise on the latter; pers. comm., George Taylor, formerly of York Textile Research Associates; and see CARDON and DU CHATENET 1990: 320.
38 For refs see BALFOUR-PAUL 1997: 86.
39 One such for wool is a seventeenth-century manuscript in the British Library, Sloane 3275, f. 61.
40 PLOWRIGHT 1903: 105; GRIERSON 1986: 214; and MAIRET 1916: 69.
41 PENLEY 1988: 141.
42 DE THEVENOT, 1686, *The Travels of M. de Thevenot into the Levant* Part 2, London (reprinted in Farnborough, Hants, in 1971): 34.
43 EDELSTEIN and BORGHETTY 1969: 166, 169.
44 For refs see BALFOUR-PAUL 1997: 86–7, also 'Nothing to hide', *Chemistry in Britain*, April 1998: 40.
45 For superstitions related to red dyes see RUMPHIUS 1750: 217; BARNES Ruth, 'East Flores Regency', in HAMILTON 1994: 189; and MOEYES 1993: 40.
46 BEMISS 1973: 138–140.
47 GOLVIN L., 1949, *Les Tissages Decorés d'El Djem*, Tunis: 78–9.
48 HERINGA 1989: 116.
49 HURRY 1930: 39–40.
50 MYERS Diana, 1990, 'Dyeing in the Himalayas', *Dyes from Nature*: 13.
51 HERINGA 1989: 115–16.
52 HOSKINS 1989: 151–4.
53 LARSEN *et al.*, 1976: 164; BÜHLER 1948: 2486; and WARMING and GAWORSKI 1981: 67.
54 NIESSEN 1985: 142; and MOEYES 1993: 40.
55 Ana Roquero, unpublished paper, 'History of Science' conference, Oxford University, January 1996.
56 HERINGA 1989: 116.
57 MAXWELL Robyn J. in OEI (ed.) 1985: 145.
58 GEIRNAERT Danielle, 'Textiles of West Sumba', in GITTINGER 1989: 69.

59 CARDON (n. 12) in CARDON *et al.*, in press.
60 Writer's fieldwork and see BRANDON 1986: 52.
61 KAUFMAN Glen, 1989, 'Mood Indigo: the Creative Genius of Hiroyuki Shindo', *Surface Design Journal*, Winter: 36.
62 RUMPHIUS 1750: 223.
63 ERB 1994 (n. 29): 206.
64 *Teinture, expression de la tradition Afrique noire* 1982: 10.
65 WENGER and BEIER 1957: 225.
66 BARBOUR and SIMMONDS 1971: 22.
67 VARADARAJAN 1985: 67.
68 BALFOUR-PAUL 1997: 89.
69 ANON 1905: 319–20.
70 CARDON 1992b: 1–14; and her article in CARDON *et al.*, in press; and see BEMISS 1973: 5–11.
71 See, for example, the 1708 recipes reprinted in *Ars Textrina*, 1990, Vol. 14: 93–9. I am grateful to Joan Thirsk for sending me the French text of a 1640s indigo recipe in the British Library, Sloane MS 2079, f. V.
72 HUMMEL 1890: and writer's fieldwork in Hungary.
73 LAWRIE 1967. See also BRUNELLO 1973: 221–74.
74 BANCROFT 1813.
75 PERSOZ 1846.
76 Walter Crum (d. 1867), chemist and calico printer (who analysed indigo), wrote 'On the manner in which Cotton unites with Colouring Matter', *Philosophical Magazine*, 24 (1844): 241–6.
77 BANCROFT 1813, Vol. 1: 165.
78 HUMMEL 1890: 297–304; and KNECHT, RAWSON and LOWENTHAL 1941, Vol. 1: 315–20.
79 See *JSCI*, 1899, Vol. XVIII: 451–7, for discussions about its composition; and see *JSDC*, 1907, Vol. XXIII, No. 2: 36–42.
80 See HORSFALL and LAWRIE 1949: 317.
81 Information supplied by Zeneca. See also PONTING 1980: 103–10; and BRUNELLO 1973: 297.
82 CARDON 1992b: 13–14.
83 HENDERSON 1950 (n. 2): 77.
84 GRIERSON 1986: 142.
85 PONTING 1976: 77 and WILLS 1970: 15.
86 KNECHT, RAWSON and LOWENTHAL 1941, Vol. 1: 323–4.
87 HURRY 1930: 39; and writer's fieldwork in Devonshire woollen mills.
88 HORSFALL and LAWRIE 1949: 311–12. See also HUMMEL 1890: 306–7.
89 Ibid.: 308; KNECHT, RAWSON and LOWENTHAL 1941, Vol. 1: 324; and VAN LAER 1874: 108–9.
90 For recipes see MAIRET 1916: 63–75; BOSENCE 1985: 115–120; and SANDBERG 1989: 113–45. And for Indian recipes see MOHANTY *et al.*, 1987: 27–37, 91–6, 158–61, 189.
91 BOSENCE 1985: 118.
92 HUMMEL 1890: 301.
93 Zeneca literature. I am grateful to Jurek Mencel, former indigo consultant to ICI, for taking me to meet Mike Quinnin at the Den-M-nit factory near Manchester.
94 Writer's fieldwork in Hungary and Austria in 1995. I am grateful to Regina Hofmann for taking me to visit Josef Koó, Austria's last blue-printer. See also HALLER 1951: 3080.
95 USHIDA Satoshi, 'Red pigments in Japanese natural indigo dye' in CARDON *et al.*, in press. Jan Wouters of the Koninklijk Instituut voor het Kunstpatrimonium in Brusssels usually finds less indirubin in European-dyed samples than in samples from other parts of the world.
96 *JSCI*, 1899, Vol. XVIII: 268; and also 1902, Vol. XXI: 222–4.
97 KNECHT, RAWSON and LOWENTHAL 1941: Vol. 1: 311–13 and Vol. 2: 827–8.
98 MOHANTY *et al.*, 1987: 98; and see CRACE-CALVERT 1876: 152–4.
99 DENHAM D. and CLAPPERTON H., 1826, *Narrative of Travels and Discoveries in Central Africa in the Years 1822, 1823 and 1824*, London: 61.
100 *JSCI*, 1899, Vol. XVIII: 1122.
101 DE NEGRI 1966: 98.
102 BALFOUR-PAUL 1997: 91–2, 95–6, 98–9.
103 LAMB 1981: 84.
104 DENHAM and CLAPPERTON 1826 (n. 99): 61.
105 Ibid.: 246.
106 BALFOUR-PAUL 1997: 95.
107 Information suplied by Jurek Mencel (n. 93).
108 HORSFALL and LAWRIE 1949: 315.
109 Writer's fieldwork in China in 1993 and in Senegal, Mauretania and Mali in 1997.
110 PICTON John, 1995, *The Art of African Textiles*, London: 12.
111 BALFOUR-PAUL 1996; and ROSSI 1990: 22–3.
112 PRANCE Ghillean Tolmie and Anne E., 1993, *Bark*, Kew (Royal Botanic Gardens): 26.
113 Pers. comm., Sandra Niessen, 1998.
114 HALPINE Susana A., 1995, 'An Investigation of Artists' Materials Using Amino Acid Analysis', *Conservation Research*, Studies in the History of Art, Monograph Series II, Washington DC: 61.
115 ROSSI 1990: 23; see also O'CONNOR Deryn, 1994, *Miao Costumes*, Farnham: 49.
116 SAYER 1985: 136.
117 See BALFOUR-PAUL 1997: 70–75 for sources.
118 BAKER Patricia L., 1995, *Islamic Textiles*, London: 20, 29, 86–7, 108–10; BALFOUR-PAUL 1997: 71–5; and see BRUNELLO 1973: 147–56.
119 See BALFOUR-PAUL 1997: 75–7 for sources.
120 BECKMAN 1846: 272–3.
121 Ahmad al-Ruqayhi (d. 1749) quoted in SERJEANT R.B. and LEWCOCK R., 1983, *San'a - An Arabian Islamic City*, London: 192a, n. 232.
122 FITZGIBBON and HALE 1997: 14, 137; and HARVEY Janet, 1996, *Traditional Textiles of Central Asia*, London: 61.
123 ISSAWI Charles (ed.), 1971, *The Economic History of Iran 1800–1914*, Chicago: 279–80 (quoting Mirza Husain's *Jughrafiyai-Isfahan*, begun in 1877).
124 SWANSON Heather, 1989, *Medieval Artisans*, Oxford: 42–3, 123–4, 133; KOWALESKI Maryanne, 1995, *Local Markets and Regional Trade in Medieval Exeter*, Cambridge: 155; and HURRY 1930: 135, 156.
125 GRIERSON 1986: 206.
126 Ibid.: 104–8; and HURRY 1930: 151–2.
127 BRUNELLO 1973: 152; and see SCHREBER 1752: 86–7.
128 BRUNELLO 1973: 156–8; CARDON 1992a; and see REBORA 1970.

129 REBORA 1970: 91.
130 HURRY 1930: 157.
131 EDELSTEIN and BORGHETTY 1969. The Sidney M. Edelstein Collection in the Jewish National and University Library in Jerusalem contains many documents relating to dyeing, including a 1548 edition of ROSETTI.
132 BRUNELLO 1973: 181–95.
133 Ibid.: 203–5.
134 Ibid.: 223–8; and MCLAREN 1986: 60.
135 HELLOT 1750:48–115, 139–55.
136 BEMISS 1973: 105.
137 *JSDC*, 1889, Vol. XIV: 134.
138 BÜHLER Alfred and FISCHER Eberhard, 1977, *Clamp Resist Dyeing of Fabrics*, Ahmedabad: 7.
139 VARADARAJAN 1985: 67; and writer's fieldwork in India, 1994.
140 BOSER-SARIVAXÉVANIS 1969: 153–6 and 190–93. Her theories were largely based on comparisons of West African textiles, most in the Basel Museum of Ethnography, the British Museum and the Musée de l'Homme in Paris. Also the present writer's fieldwork in 1997.
141 BRAVMAN R.A., 1974, *Islam and Tribal Art in West Africa*, Cambridge: 11, 883.
142 LAMB 1981: 26–8.
143 BRETT-SMITH 1990/1: 165.
144 REINHARDT Loretta, 1976, 'Mrs. Kadiato Kamara: an Expert Dyer in Sierra Leone', *Fieldiana: Anthropology*, Vol. 66, No. 2: 14.
145 PONTING 1980: 115.
146 HITCHCOCK 1991: 82–3; GILLOW 1995: 87.
147 HAMILTON 1994: 62.
148 HOSKINS 1989: 143; HERINGA 1989: 115.
149 MOEYES 1993: 41, 51.
150 HERINGA 1989: 115.
151 BRANDON 1986: 45.

Chapter Six

Opening quote: Pers. comm., Bobbie Cox, weaver, Devon, 1998.

1 FUKUMOTO Shigeki, 1993, 'Japan and the Art of Dyeing – The Japanese Dislike of Paint', *Ecotimes*, September: 13.
2 BARNES 1997, Vol. 1: 52–61. See also GITTINGER 1982: 31–7.
3 LARSEN *et al.*, 1976.
4 BÜHLER 1972. See also SEILER-BALDINGER 1994.
5 STERLING BENJAMIN 1996: Appendix.
6 Ibid: 174–91.
7 LARSEN *et al.*, 1976: 84; and STERLING BENJAMIN 1996: 57–74.
8 STEINMANN 1947: 2097.
9 FORMAN Bedrich, 1988, *Indonesian Batik and Ikat*, London: 29–30.
10 LARSEN *et al.* 1972: 77, 87–9; GILLOW 1995: 41–72; HITCHCOCK 1991: 83–97; GITTINGER 1990: 115–33; STEINMANN 1947; WARMING and GAWORSKI 1981; and GEIRNAERT Danielle C. and HERINGA Rens, 1989, *The A.E.D.T.A. Batik Collection*, Paris.
11 RAFFLES 1994, Vol. 1: 169–70 and plate 1. A collection of Raffles' batiks was acquired by the British Museum in 1859.
12 STEINMANN 1947: 2094.
13 LU PU 1981: 30.
14 Ibid. 16; ROSSI Gail, 1986, 'Laran Artists from the Mountains of China's Guizhou Province', *Surface Design Journal*, 10 No. 3: 18–22, and No. 4: 20–22; BALFOUR-PAUL 1996; and LARSEN *et al.*, 1976: 104–8.
15 STEINMANN 1949: 104–28; BUDOT E., 1994, 'Minority Costumes and Textiles of southwestern China', *Orientations*, Feb.: 59–61; O'CONNOR Deryn, 1994, *Miao Costumes*, Farnham (catalogue, West Surrey College of Art and Design); and LEWIS 1994: 108–9, 135.
16 ROSSI 1990: 23.
17 Writer's fieldwork, Erfurt in 1992, Hungary and Austria in 1995.
18 CHAI FEI *et al.*, 1956; LARSEN *et al.*, 1976: 108–9.
19 STEINMANN 1947: 2096; LARSEN *et al.*, 1976: 87.
20 Writer's fieldwork, 1993; LARSEN *et al.*, 1976: 111–12; BRANDON 1986: 29–37; and JACKSON 1997: 85–105.
21 DANIELS 1938: 126.
22 Pers. comm., the late Nancy Stanfield, 1990, and Nike Olaniyi Davies, 1997/8. See also WENGER and BEIER, 1957: 208–25; POLAKOFF 1982: 62–9; and BARBOUR and SIMMONDS 1971: 12–18.
23 POLAKOFF 1982: 72–82; WHITTLE Anita, 1998, 'Tie-Dye and Batik', *Journal for Weavers, Spinners and Dyers*, No. 185: 27–29; and LARSEN *et al.*, 1976: 82.
24 ZELTNER, FR. de. 1910, 'Tissus Africains à dessins réservés ou décolorés', *Bulletin et Mémoire de la Société d'Anthropologie de Paris*, Series 6, No. 1: 225–7; and WAHLMAN and CHUTA 1979: 455–8.
25 For fuller details see CORDWELL Justine M., 1979, 'The Use of Printed Batiks by Africans', in CORDWELL J.M. and SCHWARZ R.A. (eds), *The Fabrics of Culture*, New York; KROESE W.T., 1976, *The Origin of the Wax Block Prints on the Coast of W. Africa*, Hengelo; and PICTON John, 1995, 'Technology, Tradition and Lurex: the Art of Textiles in Africa', in PICTON John (ed.), *The Art of African Textiles*, London: 24–32.
26 I am grateful to G. Harrop, general manager of ABC, for providing information.
27 ROQUES 1678 (unpublished).
28 Writer's fieldwork in Dhamadkha, Kutch, in 1994. And see BARNES 1997: 52–61.
29 BILGRAMI 1990.
30 Writer's fieldwork, 1994. And see Mohanty B.J. and Mohanty J.P. 1983, *Block Printing and Dyeing of Bagru, Rajasthan*, Ahmedabad: 10, 13.
31 MOHANTY *et al.*, 1987: 100–144.
32 VARADARAJAN Lotika, 1982, *South Indian Traditions of Kalamkari*, Ahmedabad; and see DOSHI Sarya, 1979, 'Dazzling to the Eye': *Pichhavais* from the Deccan', *Homage to Kalamkari*, Bombay: 67–72.
33 BEAULIEU 1731–2 (unpublished).
34 SCHWARTZ 1953; and IRWIN and BRETT 1970: 7–8 and 36–58 (Appendices A–C with commentaries by P.R. Schwartz).
35 IRWIN and BRETT 1970: 5.
36 O'CONNOR and GRANGER-TAYLOR 1982: 17.
37 KATZENBERG 1973: 66; and VYDRA 1954: 14–15.
38 FLOUD 1960: 345–6; PETTIT 1974: 106–22, 195–7; BERTHOLLET and URE 1824, Vol. 2: 77–8; and PERSOZ 1846, Vol. 3: 55–6 and Vol. 4: 333–44.

39 FLOUD 1960: 346.
40 O'NEILL Charles, 1878, *The Practice and Principles of Calico Printing*, London, Vol 2: 291–92; SCHWARTZ 1953: 65–8. See also PETTIT 1974: 61; and BIEHN Michel, 1987, *En Jupon Piqué et Robe d'Indienne*, Marseilles: 26.
41 FLOUD 1960: 346–7; KATZENBERG 1973: 30–31.
42 SCHWARTZ 1953: 63–79; FLOUD 1960: 346; O'NEILL Charles, 1862, *A Dictionary of Calico Printing and Dyeing*, London: 124; BERTHOLLET and URE 1824, Vol. 2: 78–9; BRÉDIF 1989: 128–42; and CRACE-CALVERT 1876: 185–8.
43 PETTIT 1974: 83–93, 123–34; and for the development of printing machinery see TRAVIS Anthony, 1993, *From Turkey Red to Tyrian Purple*, Jerusalem.
44 FLOUD 1960: 348; PETTIT 1974; KATZENBERG 1973; and MONTGOMERY 1970.
45 HALLER 1951: 3087; *Indigo rein*, 1907: 125–60; and see BERTHOLLET and URE 1824: 81.
46 Information from ICI.
47 VYDRA 1954.
48 Writer's fieldwork in Erfurt, Germany, in 1992, and in Hungary and Austria in 1995. For Slovakia see VYDRA 1954; also BRASSINGTON 1987. See also PIROCH 1988: 63–124 (and in *Piecework* 1994: 54–60); DOMONKOS 1981; SANDBERG 1989: 48–56; WALRAVERS 1993; and BACHMANN Manfred and REITZ Günter, 1962, *Der Blaudruck*, Leipzig.
49 *Indigo rein* 1907: 204–10; and see VYDRA 1954: 35 for one basic recipe.
50 HALLER 1951: 3084.
51 *Piecework* 1994: 57; and BRASSINGTON 1987: 19.
52 BRÉDIF 1989: 48–9; and PETTIT 1974: 94–5.
53 PARNELL, Edward Andrew, 1860, *A practical treatise in Dyeing and Calico-Printing*, New York: 88; MONTGOMERY 1970: 175; and HALLER 1951: 3084–6.
54 See *Indigo rein* 1907: 171–93.
55 I am grateful to June Morris for the above information, supplemented by information and examples acquired in Capetown by Finella Balfour-Paul.
56 See KELVIN Norman (ed.), 1984, *The Collected Letters of William Morris*, Princeton, Vol. I.
57 Now in the Berger Collection, California. I am grateful to the Bergers for allowing me to study the dyebook when it was on loan in London.
58 MACCARTHY 1994: 348–61, 444; O'CONNOR Deryn, 1981, 'Red and Blue', in *William Morris and Kelmscott*, London (Design Council): 107–15; and PARRY 1983: 49–54.
59 WHITFIELD and FARRER 1990: 112.
60 BÜHLER Alfred and FISCHER Eberhard, 1977, *Clamp Resist Dyeing of Fabrics*, Ahmedabad.
61 BÜHLER 1972, Vol. I: 113–28; and LARSEN *et al.*, 1976: 129–230.
62 FORMAN 1988 (n. 9): 8.
63 See LARSEN *et al.*, 1976: 130, 152–3, 160–63, 224–7; and FITZ GIBBON and HALE 1997: Ch. 4.
64 GILLOW 1995: 83–4; HITCHCOCK 1991: 73–83; GITTINGER 1989 and 1990; WARMING and GAWORSKI 1981: 77–100; HAMILTON (ed.) 1994: 70–73.
65 Writer's fieldwork, 1992; and see CONWAY Susan, 1992, *Thai Textiles*, London: 20, 74–5, 78, 93.
66 TOMITA Jun and Noriko, 1982, *Japanese Ikat Weaving*, London: 7, 70–77; LARSEN *et al.* 1976: 210–15; and JACKSON 1997: 72–85.
67 BALFOUR-PAUL 1997: 28, 47, 126.
68 Writer's fieldwork in Syria, 1985; and see MOSER Reinhard-Johannes, 1974, *Die Ikattechnik in Aleppo*,
69 See FITZ GIBBON and HALE 1997 for examples.
70 LARSEN *et al.*, 1976: 136–7; LAMB and HOLMES 1980: 43, 186, 198, 228; and PICTON and MACK 1989: 38, 108–9.
71 MACK J., 1987, 'Weaving, Women and the Ancestors in Madagascar', *Indonesia Circle*, Vol. 42: 76–91.
72 SCHEVILL *et al.*, 1991, Chs 13–15; and pers. comm., Ann Hecht, Ana Roquero and Cati Ramsay, 1997. See also STUART Laura E., 1980 (reprint), *The McDougall Collection of Indian Textiles from Guatemala and Mexico*, Oxford: 49–52, 76–85; PENLEY 1988: 140–61; OSBORNE Lilly de Jongh, 1965, *Indian Crafts of Guatemala and El Salvador*, Oklahoma: 43–9; and SAYER 1985: 148–9, 188.
73 WADA *et al.*, 1983: 7; LARSEN *et al.* 1976: 27–73. The journal *Textileforum* devoted its 1993/1 issue to *shibori* worldwide.
74 WADA *et al.*, 1983.
75 BOSER-SARIVAXÉVANIS 1975: 320–21.
76 WAHLMAN and CHUTA 1979: 451–60; writer's fieldwork; and see WHITTLE 1998 (n. 23): 27–9.
77 *Teinture, expression de la tradition Afrique noire*, 1982; BOSER-SARIVAXÉVANIS 1969; POLAKOFF 1982: 15–54; and de LESTRANGE Monique, 1950. 'Les Sarankolé de Baydar (techniques de teinturiers)', *Études Guinéennes* (Conakry): 17–27.
78 Writer's fieldwork, 1997; and ZELTNER 1910 (n. 24): 225–7.
79 IDIENS Dale, 1980, 'An Introduction to Traditional African Weaving and Textiles', *Textile History*, Vol. 11: 15; and LAMB 1981: 29–30.
80 Writer's visit to Hiroyuki Shindo in 1993; and see KAUFMAN Glen, 1989, 'Mood Indigo', *Surface Design Journal*, Winter issue: 6–7.
81 PICTON 1980: 63–88; PICTON and MACK 1989; LAMB and HOLMES 1980; LAMB 1981 and 1984.
82 IMPERATO Pascal James, 1974, 'Bamana and Maninka Covers and Blankets', *African Arts*, Vol.VII, No. 3: 56–71; and his 'Blankets and Covers from the Niger Bend' in Vol.XII, No. 4: 38–43.
83 See BAKER Muriel and LUNT Margaret, 1978, *Blue and White – The Cotton Embroideries of Rural China*, London; and 'A Family Affair', *Hali*, June 1992, 63: 99, an exhibition review of the fine collection held at the Ashmolean Museum, Oxford.
84 WEIR 1989.
85 Now in the City of Manchester Art Galleries.
86 Ashmolean Museum, Oxford, and the Southampton Art Gallery. See also BALFOUR-PAUL 1997: 147.
87 LEWIS 1984; 76, 138, 145, 180, 204–7, 214; and BALFOUR-PAUL 1996: 99.
88 Writer's fieldwork in Japan, 1993; and MENDE and REIKO 1991. See also JACKSON 1997: 108–11; *Piecework* 1994: 22–9; and OKIKUBO, 1993, *Kogin and Shashiko Stitch*, Kyoto.

Chapter Seven

Opening quote: PHILLIPS Henry, 1822 (2nd edn), *History of Cultivated Vegetables*, London: 288.

1 BEMISS 1973: 108.
2 BALFOUR-PAUL 1997: 149–50.
3 VISSER Leontine E., 'Foreign Textiles in Sahu Culture', in GITTINGER 1989: 88, plate 10.
4 HERINGA 1989: 108–9.
5 MAXWELL 1990: 98.
6 BARNES Ruth, 'The Bridewealth Cloth of Lamalera, Lembata', in GITTINGER 1989: 49. And see BALFOUR-PAUL 1997: 156.
7 WARMING and GAWORSKI 1981: 108–14; GILLOW 1995: 86; and GITTINGER 1990: 146–49.
8 GITTINGER 1979; and MAXWELL 1990.
9 MAXWELL 1990: 407.
10 LANE Edward William, 1895, *An Account of the Manners and Customs of the Modern Egyptians*, London: 508, 520; KLUNZINGER C.B., 1878, *Upper Egypt: its People and Products*, London: 200; BLACKMAN W.S., 1927, *The Fellahin of Upper Egypt*, London: 60, 122, 264–5; and BALFOUR-PAUL 1997: 167–9.
11 WEIR 1989: 105, 148–50, 178, 185, 259.
12 Writer's fieldwork, 1985.
13 Information from Joss Graham, oriental textile specialist.
14 BOSCH Gulnar, CARSWELL John and PETHERBRIDGE Guy, 1981, *Islamic Bindings and Bookmaking*, Chicago (Oriental Institute exhibition catalogue): 34. Also information supplied by the late Don Baker, paper conservator.
15 HOSKINS 1989: 144–5, 168–70.
16 FITZ GIBBON and HALE 1997: 14, 232–3 and plate 125.
17 MAXWELL 1990: 143–4.
18 GITTINGER 1979: 31–2, 218; GAVIN 1996; and pers. comm., Traude Gavin, 1997.
19 MAXWELL 1990: 102–3.
20 HADDON Alfred C. and START Laura E., 1936 (reprint 1982, Bedford), *Iban or Sea Dayak Fabrics and their Patterns*, Cambridge: 57, 74–5, 84.
21 FRASER-LU Sylvia, in HARRIS 1993: 160–61; and MAXWELL 1990: 78.
22 GITTINGER 1990: 148.
23 CHRISTIE, in press, and CHRISTIE Jan Wisseman, 1993, 'Text and textiles in 'Medieval' Java', *Bulletin de l'École Française d'Extrême-Orient*, Vol. 80/1, Paris: 181–211.
24 HERINGA 1989: 128.
25 Ibid: 129 and plates 19 and 20. See also GILLOW 1995 plate 64.
26 MAXWELL 1990: 134–5.
27 HERINGA Rens and Harmen C., 1996, in *Fabrics of Enchantment: Batik from the North Coast of Java* (intro. GLUCKMAN Dale Carolyn), Los Angeles (County Museum of Art catalogue): 200.
28 ACHJADI Judy, 'Batiks in the Central Javanese Wedding Ceremony', in GITTINGER 1989: 155–56; and MAXWELL 1990: 134–5.
29 GITTINGER Mattiebelle in KAHLENBURG 1977: 36, and see 27–9; and her Ph.D dissertation, 'A Study of the Ship Cloths of South Sumatra: Their Design and Usage', Columbia University, New York, 1972.
30 NIESSEN 1985: 137–42; GILLOW 1995: 74–5; WARMING and GAWORSKI 1981: 97; and THEISON 1982: 59–94.
31 MAXWELL 1990: 20.
32 HAMILTON (ed.) 1994: 61–3, 104, 262.
33 GITTINGER 1979: 38.
34 WARMING and GAWORSKI 1981: 84–5, 99–100; and GITTINGER 1990: 184–91.
35 FOX James J., 1973, 'On Bad Death and the Left Hand: A Study of Rotinese Symbolic Inversions', in NEEDHAM Rodney (ed.), *Right and Left: Essays on Dual Symbolic Classification*, Chicago: 350–51, 360–64.
36 HOSKINS 1987: 149, and see 141.
37 BOND 1996 (unpublished). And see BRUNELLO 1973: 29.
38 IVENS Walter G., 1930, *The Island Builders of the Pacific*, London: 100, 122.
39 VOLLMER John E., 1980, *Five Colours of the Universe*, Edmonton, (Edmonton Art Gallery); and his 'Het traditionele begrip van de kleur blauw in China' in OEI (ed.) 1985: 171–6.
40 *Sublime Indigo* 1987: 176–81; and see WILSON 1996.
41 BETHE 1984: 65.
42 John Picton of SOAS, London University, carried out research here in the 1960s. See PICTON 1980: 63–88, especially 72–7, for the following information; also PICTON and MACK 1989: 73, 75, 77.
43 PICTON John, 1997, 'Cloth and the Corpse in Ebira', *Sacred and Ceremonial Textiles*, Proceeedings of the Fifth Biennial Symposium of the Textile Society of America, Chicago 1996: 256.
44 PICTON 1980: 74–6; PICTON and MACK 1989: 72–3; LAMB and HOLMES 1980: 226.
45 LAMB and HOLMES 1980: 79.
46 LAMB and HOLMES 1980: 197–8, and see 51, 168, 205.
47 DE NEGRI Eve, 1962, 'Yoruba Women's Costume', *Nigeria Magazine*, March, 72: 10.
48 LAMB 1981: 12, 19–49; and POLAKOFF 1982: 33, 41–8.
49 *African Textiles* 1991 (text by MACK John and SPRING Christopher): 166, plate 105.
50 Writer's fieldwork, 1997.
51 BRETT-SMITH Sarah 1990/91, 'Empty Space: the architecture of Dogon cloth ', *Res*, 19/20: 162–78.
52 Writer's fieldwork, 1997.
53 GARDI Bernard, 1985, *Ein Markt wie Mopti*, Basel (Basler Beiträge zur Ethnologie, Band 25): 206–7, 219.
54 POLAKOFF 1982: 92–3, 98.
55 KWAMI Atta, 1995, 'Textile Design in Ghana', *The Art of African Textiles – Technology, Tradition and Lurex*, London: 43; and Kathleen E., 1997, 'Mourning and memory: Factory-Printed Textiles and the Baule of Côte d'Ivoire', *Sacred and Ceremonial Textiles*, Proceeedings of the Fifth Biennial Symposium of The Textile Socity of America, Chicago 1996: 224–5.
56 Vanessa and Andrés, 1994, 'From the Infinite Blue', in *The 1994 Hali Annual*, London: 165–79.
57 DUYVETTER J., 'Rouw in de Nederlandse streekdrachten', in OEI 1985: 45–8.

58 VYDRA 1954: 45.
59 Writer's fieldwork, 1995; and BRASSINGTON 1987: 19, 25.
60 BEAGLE 1975: 39–41.
61 HYMAN Virginia Dulany and HU William C.C., 1982. *Carpets of China and its Border Regions*, Ann Arbor, Michigan: 73, 128.
62 SHEAF Colin D., 1996, 'China for Europe', in *Silk and Stone, the Art of Asia*, London: 77; and see CARSWELL John, 1995, 'White and Blue', *Asian Art*, London: 154–67.
63 Writer's fieldwork in Hungary and Austria, 1995.
64 STARK George, 'A social-economic study of the woad-blue sources of the Blue Monday', in CARDON *et al.*, in press.
65 HURRY 1930: 48.
66 SALZMAN 1923: 211–14; see also HURRY 1930: 49–50.
67 Henry IV, Part 2, Act V, Scene IV.
68 Information from Wilf Joint, an employee at the Buckfastleigh woollen mills in Devon from 1933 to 1974.
69 PONTING 1976: 87.
70 LEGGETT 1944: 30–31.
71 *JSCI*, 1899, Vol. XVIII: 718.
72 HERINGA 1989: 127.
73 HAMILTON 1994: 104.
74 RATHBUN William J., 1993, *Beyond the Tanabata Bridge, Traditional Japanese Textiles*, London: 32.
75 BRANDON 1986; JACKSON 1997; and *Sublime Indigo* 1985: 186–7.
76 LEWIS 1984; Zhao Yuchi and Kuang Shizhao (eds), 1985, *Clothings and Ornaments of China's Miao People*, Beijing. See also BUDOT Eric, 1994, 'Minority Costumes and Textiles of Southwestern China', *Orientations*, Feb.: 59–66.; O'CONNOR Deryn, 1994, *Miao Costumes*, Farnham (catalogue, West Surrey College of Art and Design); and BALFOUR-PAUL 1996: 94–9.
77 HYMAN Virginia Dulany and HU William C.C., 1982, *Carpets of China and its Border Regions*, Michigan: 114.
78 COLE Thomas J., 1990, 'A tribal tradition', *Hali*, February, 49: 16–29.
79 TANAVOLI Parviz, 1992, '*Shisha Derma* – Iranian 'Black and White' Textiles', *Hali* 63: 84–89. See also MUSHKAT Fred, 1996, 'Persian warp-faced nomadic bands', *Hali* 84: 78–83.
80 For a much fuller account see BALFOUR-PAUL 1997: 117–54.
81 BALFOUR-PAUL 1997: 129–30; and BAKER Patricia L., 1995, *Islamic Textiles*, London: 16–17, 20.
82 Cited by BALDRY John, 1982, *Textiles in Yemen*, British Museum Occasional Paper No. 27, London: 49.
83 DE LANDBERG Carlo, 1901–13, *Etudes sur les Dialectes de l'Arabie Méridionale*, Leiden: 419.
84 LAMB and HOLMES 1980: 93.
85 See, for example, PICTON and MACK 1989: 194–5.
86 LEO AFRICANUS, *Description de L'Afrique* (trans. Epaulard A. 1956), Paris: 35.
87 BARBOT 1732.
88 ROBERTS Richard, 1984, 'Women's Work and Women's Property: Household Social Relations in the Maraka Textile Industry of the Nineteenth Century', *Comparative Studies in Society and History*, Vol. 26, No. 2: 235.
89 LAMB and HOLMES 1980: 92; and see Hausa dictionary definitions of *shunin batta/ban shuni*.
90 See, for example, *Adire, Resist-Dyed Cloths of the Yoruba*, exhibition booklet, National Museum of African Art, Smithsonian Institution, Washington DC, 1997.
91 DE NEGRI 1962 (n. 47): 4.
92 Pers. comm., Nike Olaniyi-Davies, 1998.
93 BALFOUR-PAUL 1997: 46.
94 See SCHEVILL *et al.* 1991: 309–78 (for essays by DAVIS Virginia, MILLER Laura Martin, and CARLSEN Robert S. and WENGER David A.).
95 OSBORNE Lilly de Jongh, 1965, *Indian Crafts of Guatemala and El Salvador*, Oklahoma: 44.
96 SAYER 1985: 143–4 (and pers. comm.); SAYER Chloë and BATEMAN Penny, in HARRIS (ed.) 1993: 276–8; and SKIRVIN Susan, 1990, 'Mayo Indigo', *Dyes from Nature*, Brooklyn Botanic Garden Record, Vol. 46, No. 2, Summer: 40–41.
97 THOMPSON Jon, 1995, 'Shellfish Purple: The Use of *Purpura patula pansa* on the Pacific Coast of Mexico', *DHA* 13: 3–6. See also SAYER 1985: 145.
98 *Blu Blue-Jeans*, Genoa exh. cat., Milan 1989.
199 GINOUVEZ Olivier, 1993, 'Les fouilles de la Z.A.C. des Halles à Nîmes (Gard)', *Bulletin de l'Ecole Antique de Nîmes*, Suppl. 1: 199–209.
100 I am grateful to Lynn Downey of Levi Strauss & Co., and Zeneca, makers of indigo for BASF company, for providing much of the following information.
101 *Lands' End* clothing catalogue, March 1995.
102 WHITFIELD and FARRER 1990: 114–15, plate 89.
103 Information from Den-M-nit, 'The Real Indigo Company', Burnley.

Chapter Eight

Opening Quote: BORRADAILE, etc. (see n.16): 59 for ref.

1 GUINEAU Bernard and DULIN Laurent, 'The Colours of Indigo in Painted Manuscripts 9th–11th Centuries', in CARDON *et al.*, in press; HARLEY 1982: 70, 76–88; and CENNINO CENNINI 1960: 28–9, 32.
2 Confirmed by Raman spectrometry identification – see GUINEAU and DULIN (n. 1). See also 'Le secret du bleu Maya', 1996, *Sciences et Avenir*, No. 595: 90.
3 *Dye Plants and Dyeing* 1964: 83–5.
4 BIRD Junius B., 1973, *Peruvian Paintings by Unknown Artists 800BC to 1700AD*, New York; and REID James, 1990, 'Magic Realism – Painted Textiles from the Central Coast of Peru', *Hali*, 54, December 1990: 122–31.
5 See p.23. See also BAILEY KC. (ed.), 1932, *The Elder Pliny's Chapters on Chemical Subjects*, London, Part 2: 86–89 and 216–24; and BECKMAN 1846: 258–71.
6 STILLMAN J.M., 1960, *The Story of Alchemy and Early Chemistry*, New York, (1st edn, London 1924): Ch. 2.
7 ZERDOUN BAT-YEHOUDA M., 1983, *Les Encres Noires au Moyen Age*, Paris: 18, 82–3, 138, 331.
8 BROWN Percy, 1975, *Indian Painting under the Mughals, AD 1550–AD 1750*, New York: 187–9; and DAS J.P., 1982, *Puri Paintings*, New Delhi: 9–16, 90–98.

9 Pers. comm., the late manuscript specialist Don Baker, and see his 'The Conservation of Jami al-Tawarikh by Rashid al-Din (1313), *Arts and the Islamic world*, No. 20 (Spring 1991): 33.
10 For this, and other sources, see Martin Levey, 1962, *Mediaeval Arabic Bookmaking and its Relation to Early Chemistry and Pharmacology*, Transactions of the American Philosophical Society, Vol. 52, Part 4, Philadelphia: 6–7.
11 Ibid: 9, 23, 26–7, 31–2.
12 NEEDHAM Joseph (ed.), 1985, *Science and Civilisation in China*, Vol. 5, Part 1 (by TSUEN-HSUIN Tsien): 247.
13 YING-HSING 1966: 75; and FORTUNE R., 1846, 'A Notice of the Tien-ching, or Chinese Indigo', *Royal Horticultural Society Journal*: 270.
14 POMET 1694: 154–6.
15 PARKINSON 1640: 602. PLOSS E.E., 1962, *Ein Buch von alten Farben . . .'*, Heidelberg and Berlin: 113. BL MS Sloane 122: f.70r.
16 See BORRADAILE Viola and Rosamund, 1966, *The Strasbourg Manuscript*, London: 103.
17 HARLEY 1982: 66–7.
18 Pers. comm. Jo Kirby, National Gallery, London.
19 EASTLAKE Charles Lock, 1960, *Materials of Painting of the Great Schools and Masters*, New York, Vol. 1: 120–21.
20 CENNINO CENNINI 1960: 52.
21 BORRADAILE 1966 (n. 16): 45, 55, 59, 61.
22 For examples see *Sublime Indigo* 1987: 82–97.
23 CENNINO CENNINI 1960: 29, 32, 36, 52, 55, 91, 105, 117. See also THOMPSON D.V., 1936, *The Materials of Medieval Painting*, London: 138–9, 173;
24 GUINEAU and DULIN (n. 1).
25 See, for example, GIBBS Peter J. *et al.*, 1995, 'Letter: The *in situ* identification of indigo on ancient papers', *European Mass Spectrometry* 1: 417–21; and WITHNALL Robert *et al.*, 1993, 'Non-destructive, *in situ* identification of indigo/woad and shellfish purple by Raman microscopy and visible reflectance spectroscopy', *DHA* 11: 19–24.
26 HARLEY 1982: 69–70; and EASTLAKE 1960 (n. 19): 454.
27 HARLEY 1982: 26, 70, 162–3.
28 LEHNER Sigmund (trans. MORRIS Arthur and ROBSON Herbert), 1902, *Ink Manufacture including Writing, Copying, Lithographic, Marking, Stamping and Laundry Inks*, London: 43–8,104,140–1,157–9.
29 See, for example, DANIELS Vincent, 1990, 'Anomalous Fading of Indigo-Dyed Paper', *DHA*, 8: 17–18.
30 SINCLAIR Eddie, 1995, 'Polychromy of Exeter and Salisbury Cathedrals: A preliminary Comparison', *Historical Painting Techniques, Materials and Studio Practice* (ed. WALLERT Arie), Leiden (Getty Conservation Institute conference): 105–10.
31 PLAHTER Unn, 1995, 'Colour and Pigments used in Norwegian Altar Frontals', *Norwegian Medieval Altar Frontals and Related Material* (ed. BRETSCHNEIDER Giorgi), Rome (conference papers, Olso, December 1989): 111–26.
32 ERSKINE A.M. (ed. and trans.), 1981, *The Accounts of the Fabric of Exeter Cathedral, 1279–1353*, Part 1 (1279–1326), Devon and Cornwall Record Society, New Series, Vol. 24, Torquay: 134, 142.
33 BRISTOW 1996: 14–15,160,165–7,169.
34 TINGRY P.F., 1830, *The Painter's and Colourman's Complete Guide*, London (3rd edn): 145–6.
35 OKE O.L. in BARBOUR and SIMMONDS (eds) 1971: 45.
36 FRASER-LU Sylvia, 1994, *Burmese Crafts Past and Present*, Oxford: 68, 223.
37 Pers. comm., Amit Ambalal, Ahmedabad artist and author of *Krishna as Shrinathji, Rajasthani Paintings from Nathdvara*, Ahmedabad, 1987.
38 Pers. comm., Toofan Rifai of Sarkej, near Ahmedabad.
39 RIEFSTHAL E., 1944, *Patterned Textiles in Pharaonic Egypt*, Brooklyn: 16–17.
40 For example see BOSCH Gulnar, CARSWELL John and PETHERBRIDGE Guy 1981, *Islamic Bindings and Bookmaking*, (exhibition catalogue, Oriental Institute), Chicago.
41 EDELSTEIN and BORGHETTY 1969: 159,165,169,170,176.
42 CENNINO CENNINI 1960: 11–12.
43 NEEDHAM 1985 (n. 12): 76–7.
44 Pers. comm., Beth McKillop, British Library, and Jean-Pierre Drège, 'Centre de recherche sur les manuscrits, inscriptions et documents iconographiques de Chine', Paris.
45 Pers. comm., Dr Cristina Anna Scherrer-Schaub, Lausanne.
46 Marble-paper making, popular in sixteenth-century Turkey, thereafter influenced Europe – see *Sublime Indigo* 1987: 102–7.
47 PORTER Yves (trans. by BUTANI S.), 1994, *Painters, Painting and Books: An essay on Indo-Persian Technical Literature, 12–19th Centuries*, New Delhi: 36–43.
48 BRÜCKLE I., 1993, *'the historical manufacture of blue-coloured paper'*, *Paper Conservator*, Vol. XVII: 20; and *Sublime Indigo* 1987: 77–8.
49 KRILL John, 1987, *English Artists' Paper*, London (V&A exh. cat.): 56, 61.
50 Information leaflet by Brad Sabin Hill, for exhibition of *Carta Azzura*, British Library, 1995.
51 BRÜCKLE 1993 (n. 48): 20–31; KRILL 1987 (n. 49): 90–95; and *Sublime Indigo* 1987: 98–115 *passim*.
52 KRILL 1987 (n. 49): 61.

Chapter Nine

Opening quote: Vaidya Bhagwan Dash, 1994, *Materia Medica of Tibetan Medicine*, Indian Medical Series No. 18, Delhi: 53.

1 VARLEY H., 1980, *Colour*, London: 46.
2 VARADARAJAN 1985: 65–7; and TURNER C.W., 1937, 'Colour Symbolism in India', *CR* 2: 57–61.
3 GOITEIN S.D., 1967–88, *A Mediterranean Society*, Vol. 4: 174–5.
4 BALFOUR-PAUL 1997: 156–7.
5 HOSKINS 1989: 141–73.
6 LEMMENS and WULIJARNI-SOETJIPTO 1991: 32.
7 VARADARAJAN 1985: 65; BRANDON 1986: 43; YING-HSING 1966: 78; and HYMAN Virginia Dulany and HU William C.C., 1982, *Carpets of China and its Border Regions*, Ann Arbor, Michegan: 113.
8 *Historia Naturalis*, Liber XXXV, Sect.XXVII,(46).

9 DIOSCORIDES, *Materia Medica*, Book II, Chs 107 (indigo), 215 (woad). And see HURRY 1930: 249–53.
10 IBN AL-BAYTAR, 'Traité des Simples', Vol. 23 Part I, Vol. 25 Part I and Vol. 26 Part I of LECLERC N.L., 1877–83, *Notices et Extraits des Manuscrits de la Bibliothèque Nationale*, Paris.
11 DE L'OBEL Matthioe, 1576, *Stirpium Historia*: 189–91; see also DODOENS D. Rembert (trans. LYTE Henry), 1619, *A New Herbal or Historie of Plants*, London: 48. And see BALFOUR-PAUL 1997: 159.
12 SALMON William, 1710, *The English Herbal*: 1273.
13 GERARDE John, 1597, *The Herball*: 394–5.
14 CULPEPER N., 1826, *Complete Herbal and English Physician*, Manchester (facsimile, Harvey Sales 1981): 200.
15 CARDON and du CHATENET 1990: 146.
16 PHILLIPS Henry, 1822, *History of Cultivated Vegetables*, London Vol. I: 293.
17 STUART G.A., 1979, *Chinese Materia Medica*, Taipei (original edn published in Shanghai, 1911): 218.
18 BURKILL 1935: 1237.
19 WATT John Mitchell and BREYER-BRANDWIJK Maria Gerdina, 1962 (2nd edn), *The Medicinal and Poisonous Plants of Southern and Eastern Africa*, Edinburgh and London: 611–13.
20 See KIRTAKAR K.R. and ., 1918, *Indian Medicinal Plants*, Delhi, Vol. I: 708 and 712–14.
21 ISMAIL BHAI, n.d., *Gujarati Herbal*.
22 Zeneca literature.
23 For sources on Central and South America see HASTINGS Rupert B., 1990, 'Medicinal Legumes of Mexico: Fabaceae, Papilionoideae, Part One', *Economic Botany*, Vol. 44 (3): 336–48. See also MARTINEZ Maximino, 1933, *Las Plantas Medicinales de México*, Mexico: 38–41; *Nuevo Farmacopea Mexicano*, 1952 (sixth edn), Mexico: 66–7; DOMINGUEZ Xorge A., *et al.*, 1978, 'Mexican Medicinal Plants XXXI', *Planta Medica, Journal of Medicinal Plant Research*, Vol. 34 (2): 172–5; ARNOLD 1987: 70–71,76; and *Dye Plants and Dyeing* 1964: 83–5.
24 Writer's fieldwork, 1985, and research by Gigi and Roddy Jones, 1985–9. In North Africa indigo has also been used medicinally, but there henna is the main cure-all.
25 BALFOUR-PAUL 1997: 161–2.
26 CHANG Hson-Mon and BUT Paul Pui-Hay, 1986, *Pharmacology and Applications of Chinese Materia Medica*, Singapore, Vol. I: 715.
27 MARTINEZ 1933 (n. 23): 39; *Nuevo Farmacopea Mexicana* 1952 (n. 23): 66–7; and HASTINGS 1990 (n. 23): 345.
28 KIRTAKAR and BASU 1918 (n. 20): 10.
29 WATT and BREYER-BRANDWIJK 1962 (n. 19): 611–13. See also KOKWARO J.O., 1976, *Medicinal Plants of East Africa*, Kampala: 137–8.
30 STUART 1979 (n. 17): 218.
31 BALFOUR-PAUL 1997: 161.
32 WALKER J., 1934, *Folk Medicine in Modern Egypt*, London: 45.
33 HASTINGS 1990 (n. 23): 345.
34 See under *Shuni* in BARGERY G.P., 1931, *A Hausa–English Dictionary*, London. Under the term *baba* indigo is also noted as a preventative against syphilis.
35 HOSKINS 1989: 142–52, 170.
36 Ibid.: 142.
37 BARGERY 1931 (n. 34).
38 Writer's fieldwork, 1994.
39 Articles mentioning indigo have appeared in such journals as *Economic Botany* (n. 23); *Phytochemistry*, e.g. MAIER W., SCHUMANN B. and GROGER D., 1990, 'Biosynthesis of indoxyl derivatives in *Isatis tinctoria* and *Polygonum tinctorium*' in No. 29 (3): 817–19;and the *Journal of Ethnopharmacology*, e.g. HAN J., No. 24(1): 1–17.
40 SONG Zhijun *et al.*, 1996, 'Effects of Chinese medicinal herbs on a rat model of chronic *Pseudomonas aeruginosa* lung infection', *APMIS* 104: 350–54.
41 CHANG and BUT 1986 (n. 26): 712–16.
42 Ibid.: 694–700.
43 Writer's fieldwork in China and Japan, 1993; and see, for example, TOMITA, J. and N., 1982, *Japanese Ikat Weaving*, London: 70–71.
44 WATT and BREYER-BRANDWIJK 1962 (n. 19): 612.
45 CREYAUFMÜLLER Wolfgang, 1985, 'De mythe van de Blauwe mannen', in OEI (ed.) 1985: 99–102, and BALFOUR-PAUL 1997: 151–3.
46 BALFOUR-PAUL 1997: 140–42.
47 Julius Caesar, *Commentarii de bello Gallico*, Liber V, XIV(2): *'omnes vero se Britanni vitro inficiunt, quod caeruleum efficit colorem atque hoc horridiores sunt in pugna aspectu'* ('All the Britanni spot themselves with woad [as translated], which produces a blue colour and gives them a horrible appearance in battle').
48 See PYATT F.B., BEAUMONT E.H., LACY D., MAGILTON J.R. and BUCKLAND P.C., 1991, 'Non Isatis Sed Vitrum, or The Colour of Lindow Man', *Oxford Journal of Archaeology*, 10(1): 61–73; VAN der VEEN M., HALL A. and MAY J., 1993, 'Woad and the Britons Painted Blue', *Oxford Journal of Archaeology*, 12: 367–71.
49 CORDWELL Justine M., 1979, 'The Very Human Arts of Transformation', in CORDWELL Justine M. and SCHWARTZ Ronald A. (eds), *The Fabrics of Culture*, The Hague: 68–70; and SEEFELDER 1994: 34–7.
50 DENHAM D. and CLAPPERTON H.,1826, *Narrative of Travels and Discoveries in Central Africa in the Years 1822, 1823 and 1824*, London, Section II: 61.
51 BALFOUR-PAUL 1997: 162–5.
52 STARK Freya, 1940, *A Winter in Arabia*, London (indexed version): 58.
53 See BALFOUR-PAUL 1997: 163 for refs.
54 STARK Freya, 1936, *The Southern Gates of Arabia*, London: 18.
55 FOUQUET Dr, 1899, 'Le Tatouage Médicale en Egypte dans l'Antiquité', in *Archives d'Anthropologie Criminelle* XIII: 270–79 and figs 1–25.
56 Ibid.
57 LANE Edward William, 1895, *An Account of the Manners and Customs of the Modern Egyptians*, London: 48–9.
58 *Naval Intelligence Geographical Handbook – Western Arabia and the Red Sea*, June 1946: 418.
59 e.g. Southeast Asia – see MAXWELL 1990: 109–10.
60 HERBER Jean. 1948, 'Tatoueuses Marocaines', *Hespéris*, Paris, Vol. 35: 289–97.

61 CORDWELL 1979 (n. 49): 58–60; and see BOHANNAN Paul, 1965, 'Beauty and Scarification amongst the Tiv', *Man*, Sept.: 118.
62 GITTINGER Mattiebelle, 'Sier en Symbool', in OEI (ed.) 1985: 163.
63 FOSTER 1968: 40.
64 HURRY 1930: 117.
65 SAHAÚN Bernardino de, 1985, *Historia General de las Cosas de Nueva España* (ed. Porrúa), Mexico: 699.
66 JUYNBOLL G.H.A., 1986, 'Dyeing the hair and beard in early Islam. A Hadith analytical study', *Arabica*, March, Vol. XXXIII: 49–75.
67 SAID H.M.(trans. and ed.), 1973, *Al-Biruni's Book on Pharmacy and Materia Medica*: 229.
68 CHARDIN Jean, 1927, *Travels in Persia* (ed. PENZER N.M., from the English translation of 1720), London: 216.
69 MORIER James, 1812, *A Journey through Persia, Armenia and Asia Minor to Constantinople in the years 1808 and 1809*, London: 231.
70 PORTER Robert Ker, 1821, *Travels in Georgia, Persia, Armenia etc. during the years 1817,1818,1819 and 1820*, London, Vol. 1: 231–3, and see 327–8, 356.
71 HONINGBERGER John Martin, 1852, *Thirty-Five Years in the East*, 167–8.
72 HEMNETER Ernst, 1937, 'The Castes of the Indian Dyers', *CR* 2: 54.
73 BILGRAMI 1990: 103.
74 See *shuni* in a Hausa dictionary (n. 34).
75 HOULBERG Marilyn Hammersley, 1979, 'Social Hair: Yoruba Hairstyles in Southwestern Nigeria', in CORDWELL and SCHWARTZ (n. 49): 367–73.
76 Pers. comm., John Picton. These practices may still take place.
77 IVENS Walter G., 1927, *Melanesians of the S.E. Solomons*, London: 20.
78 IBN AL-BAYTAR in LECLERC 1877–83 (n. 10).
79 KIRTAKAR and BASU 1918 (n. 20): 710.
80 WATT and BREYER-BRANDWIJK 1962 (n. 19): 612; and HASTINGS 1990 (n. 23): 345.
81 AL-GHOUL M. *et al.*, 1980, 'Al-Tibb al-sha^{c}bi', in *Nadwat al-darasat al-Umaniya* (Symposium on Omani Studies) Muscat, Vol. 5: 108.
82 BURKILL 1935, Vol. 2: 1237.
83 I am grateful to Polly Lyster, indigo dyer, for acquiring a bottle of Anoop herbal hair oil for me when she was in India in 1995.

Chapter Ten

1 FUKUMOTO Shigeki, 1993, 'Japan and the Art of Dyeing – The Aesthetics of Age', *Ecotimes*, December: 13.
2 Pers. comm., Charles Day, managing director of Henry Day and Sons Ltd., fibre reclaimers and blenders in Dewsbury, Yorkshire.
3 FARKASVÖLGYI-GOMBOS, Zsuzsa, 1985, 'Székely Flat-Woven Rugs of Eastern Transylvania', in *Oriental Carpet and Textile Studies* 1, London: 226.
4 SMITH Roy and WAGNER Sue, 1991, 'Dyes and the Environment: Is Natural Better?', *American Dyestuff Reporter* (September): 34; and pers. comm., Graham Ashton of Zeneca.
5 *Science and Technology*, 1997, 'A gene to make greener blue jeans', March: 82–3; ALDRIDGE Susan, 1997, 'Growth Industry', *New Scientist, Inside Science*, No. 105: 1–4; and *Biotechnology Means Business (BMB)*, Dept. of Trade and Industry UK status report for the textile and clothing industries, 1995: 17.
6 PADDEN A. Nikki, DILLON Vivian M., JOHN Philip EDMONDS John, COLLINS M. David and ALVAREZ Nerea, 1998. '*Clostridium* used in medieval dyeing', *Nature*, Vol. 396, 19 November: 225.
7 BECHTOLD T., BURTSCHER E., KÜHNEL G. and BOBLETER O., n.d. (after 1995), 'Electrochemical reduction processes in indigo dyeing', paper produced for the Institute for Textile Chemistry and Physics, Leopold-Franzens University, Innsbruck, Austria.
8 This system is used by Den-M-Nit, 'The Real Indigo Company Ltd', of Burnley, UK.
9 Pers. comm., Graham Ashton of Zeneca.
10 See 'Bio-Piracy', 1993, *Rafi Communique* (Rural Advancement Foundation International), Nov/Dec. See also *BMB* (n. 5): 12.
11 MESTEL Rosie, 1993, 'How blue genes could green the cotton industry', *New Scientist*, 31, July: 7.

Select Bibliography

Abbreviations

AEDTA Association pour l'Étude de la Documentation des Textiles d'Asie
CR *Ciba Review*
DHA *Dyes in History and Archaeology*
JSCI *Journal of the Society of Chemical Industry*
JSDC *Journal of the Society of Dyers and Colourists*

Published works

ADROSKO Rita J., 1971. *Natural Dyes and Home Dyeing*, New York.

AEDTA, 1990. *La teinture végétale: le bleu*, Paris.

African Textiles, 1991, exh. cat. (Japanese and English), Kyoto.

ALDEN D., 1965. 'The growth and decline of indigo production in colonial Brazil: a study in comparative economic history', *Journal of Economic History*, Vol. 35: 35–60.

ANON, 1705. *The Whole Art of Dying* [*sic*], London (trans. from German).

ARNOLD Dean, 1987. 'The Evidence for Precolumbian Indigo in the New World', *Anthropología y Técnica*: 53–83.

BAILEY K.C. (ed.), 1932. *The Elder Pliny's Chapters on Chemical Subjects*, Part 2, London.

BALFOUR-PAUL Jenny, 1987. 'Indigo – an Arab Curiosity and its Omani Variations', in PRIDHAM B.R. (ed.), *Oman: Economic, Social and Strategic Developments*, London: 79–93.

BALFOUR-PAUL Jenny, 1990. 'The Indigo Industry of the Yemen', *Arabian Studies* VIII (University of Cambridge Oriental Publication 42), Cambridge: 39–62.

BALFOUR-PAUL Jenny, 1993. 'The International Conference on Woad in Erfurt', *DHA* 11: 3–5.

BALFOUR-PAUL Jenny, 1995. 'The Woad Trade of Toulouse and the Second International Conference on Woad, Indigo and Other Natural Dyes', *DHA* 13: 77–81.

BALFOUR-PAUL Jenny, 1996. 'Guizhou Blue, The Traditional Use of Indigo in Southwest China', *Hali* 89, October: 94–9.

BALFOUR-PAUL Jenny, 1997. *Indigo in the Arab World*, London.

BALFOUR-PAUL Jenny, 1999. 'Europe's Woad Revival', *Textile Forum* 4: 26–7.

BANCROFT Edward, 1813 (2nd edn). *Experimental Researches Concerning the Philosophy of Permanent Colours. . .*, 2 Vols, London.

BARBOT John, 1732. *A Description of the Coasts of North and South-Guinea*, London (Vol. 5 of CHURCHILL A., *A Collection of Voyages and Travels*).

BARBOUR Jane and SIMMONDS Doig (eds), 1971. *Adire Cloth in Nigeria*, Ibadan.

BARNES Ruth, 1997. *Indian Block-Printed Textiles in Egypt: The Newberry Collection in the Ashmolean Museum* 2 Vols, Oxford.

BASU Kamud Vihari (ed.), 1903. *Indigo Planters and All About Them*, Calcutta.

BEAGLE Peter, 1975. *American Denim: A New Folk Art*, New York.

DE BEAUVAIS-RASEAU, 1770. *L'Art de l'indigotier*, Paris.

BECKMAN John, 1846 (4th edn). *History of Inventions, Discoveries and Origins*, Vol. 2, London.

BEMISS Elijah, 1973 (reprint). *The Dyer's Companion* (intro. by R.J. Adrosko), New York.

BENNECKENSTEIN Horst, n.d. *Waid des Thüringer Landes goldenes Blies*, Thüringer Freilichtmuseum Hohenfelden.

BENNECKENSTEIN Horst (ed.), 1988, 1989 and 1991. *Beiträge zur Waidtagung* (conference papers), Gotha (1988/9) and Pferdingsleben (1991).

BÉRARD H., 1838. *Nouveaux essais d'extraction de l'indigo du polygonum tinctorium*, Montpellier.

BERTHOLLET C.L. and URE A.B., 1824. *Elements of the Art of Dyeing* (trans. from French by Andrew Ure), 2 Vols, London.

BETHE Monica, 1984. 'Color: Dyes and Pigments', in STINCHECUM Amanda Mayer, *Kosode, 16th–19th Century Textiles from the Nomura Collection*, New York (Japan Society and Kodansha International).

BETTEY J.H., 1978. 'The Cultivation of Woad in the Salisbury Area during the Late Sixteenth and Seventeenth Centuries', *Textile History*, Vol. 9: 112–17.

BICKLEY Francis B. (ed.), 1900. *The Little Red Book of Bristol*, Vol. 2, Bristol and London.

BILGRAMI Noorjehan, 1990. *Sindh Jo Ajrak*, Karachi.

BISHOP Leander, J., 1868 (3rd edn). *A History of American Manufactures from 1680–1860*, Vol. 2, Philadelphia.

BOLLAND Rita, 1991. *Tellem Textiles*, Leiden.

BOSENCE Susan, 1985. *Hand Block Printing and Resist Dyeing*, Newton Abbot.

BOSER-SARIVAXÉVANIS Renée, 1969. *Aperçus sur la teinture à l'indigo en Afrique Occidentale*, Basel.

BOSER-SARIVAXÉVANIS Renée, 1972. *Les Tissus de'Afrique Occidentale*, Basel.

BOSER-SARIVAXÉVANIS Renée, 1975. *Recherche sur l'histoire des textiles traditionnels tissés et teints de l'Afrique Occidentale*, Basel.

BRANDON Reiko Mochinaga, 1986. *Country Textiles of Japan*, New York and Tokyo.

BRASSINGTON Linda, 1987. *Modrotlac: Indigo Country Cloths and Artefacts from Czechoslovakia*, exh. cat. (West Surrey College of Art and Design), Farnham.

BRÉDIF Josette, 1989. *Toiles de Jouy*, London.

BRETT-SMITH Sarah, 1990/91. 'Empty Space: the architecture of Dogon Cloth', *Res* 19/20: 163–78.

BRISTOW Ian C., 1996. *Interior House Painting Colours and Technology 1615–1840*, New Haven.

BRÜCKLE I., 1993. 'The historical manufacture of blue-coloured paper', *Paper Conservator*, Vol. XVII: 20–31.

BRÜGGEMANN W. and BÖHMER H., 1982. *Teppiche der Bauern und Nomaden in Anatolien*, Munich.

BRUNELLO Franco, 1973. *The Art of Dyeing in the History of Mankind*, Vicenza.

BÜHLER Alfred, 1948. 'Dyeing among Primitive Peoples', *CR* 68: 2478–2511.

BÜHLER Alfred, 1951. 'Indigo Dyeing among Primitive Races', *CR* 85: 3088–91.

BÜHLER Alfred, 1972. *Ikat, Batik, Plangi*, 3 Vols, Basel.

BURKILL I.H., 1935. *A Dictionary of the Economic Products of the Malay Peninsula*, Vol. 2, London.

CARDON Dominique and du CHATENET Gaëtan, 1990. *Guide des teintures naturelles*, Paris.

CARDON Dominique, 1992a. 'New information on the medieval woad vat', *Dyes in History and Archaeology*, 10: 22–31.

CARDON Dominique, 1992b. 'From the medieval woad-vat to the modern indigo-vat', *Beiträge zur Waidtagung*, Vol. 4/5, Part 1: 10–15 and Appendices 1 and 2.

CARDON D., MÜLLEROTT H.E. *et al.* (eds), in press. *Actes/Papers/Beiträge, Second International Symposium on 'Woad, Indigo and Other Natural Dyes: Past Present and Future 1995* (*Beiträge zur Waidtagung*, Vol. 7), Toulouse/Arnstadt.

CARDON Dominique, forthcoming. *Technologie de la draperie mediévale*, Paris.

CASTER Gilles, 1962. *Le commerce du pastel et de l'épicerie à Toulouse, de 1450 environ à 1561*, Toulouse.

CENNINO d'Andrea CENNINI, 1960. *The Craftsman's Handbook, 'Il Libro dell'Arte'* (trans. THOMPSON D.V.), New York.

CHAI FEI, HSU CHEN-PENG, CHENG SHANG-JEN and WU SHU-SHENG, 1956. *Indigo Prints of China*, Beijing.

CHAKLADAR H.C., 1960. *Papers Relating to the Cultivation of Indigo in the Presidency of Bengal*, Calcutta.

CHAKLADAR H.C., 1972. 'Fifty Years Ago: the Woes of a Class of Bengal Peasantry under European Indigo Planters', *Human Events* 6 (54): 38–46.

CHAKLADAR H.C., 1973. 'Further note on Indigo Disturbances', *Human Events* 7 (61): 25–40.

CHARPENTIER-COSSIGNY DE PALMA J.F., 1789 (reprint 1970). *Memoir, Containing an Abridged Treatise on the Cultivation and Manufacture of Indigo*, Calcutta.

CHAUDHURI K.N., 1965. *The English East India Company – The Study of an Early Joint-Stock Company 1600–1640*, London.

CHAUDHURI K.N., 1978. *The Trading World of Asia and the English East India Company 1660–1760*, Cambridge.

CHEESMAN Patricia, 1988. *Lao Textiles*, Bangkok.

CHEVREUL M.E., 1830. *Leçons de Chimie appliqué à la teinture*, Paris.

CHRISTIE Jan Wisseman, in press. 'Fibres and dyestuffs in early Java and Bali', in BINTARTI D.D., SUTABA I. Made and GHOSH Asok K. (eds), *Man, Culture and Environment in Prehistoric South and Southeast Asia*, Jakarta.

CLARK H.O. and WAILES Rex., 1935–6. 'The Preparation of Woad in England', *Transactions of the Newcomen Society*, Vol. XVI.

CLARK Robin J.H., COOKSEY Christopher J., DANIELS Marcus A.M. and WITHNALL Robert, 1993. 'Indigo, Woad and Tyrian Purple: important vat dyes from antiquity to the present', *Endeavour*, Vol. 17/4 (new series): 191–9.

COON D.L., 1976. 'Eliza Pinckney and the reintroduction of indigo culture in South Carolina', *Journal of Southern History*, Vol. 42: 61–76.

CRACE-CALVERT F., 1876 (2nd edn). *Dyeing and Calico Printing*, Manchester.

CRAWFURD John, 1820. *History of the Indian Archipelago*, Edinburgh.

CROLACHIUM H., 1555. *Isatis Herba de cultura herbae isatidis quam guadum...*, Gotha (also printed in SCHREBER 1752).

CROOKES William, 1874. *A Practical Handbook of Dyeing and Calico-Printing*, London.

DANIELS F., 1938. 'Yoruba pattern dyeing', *Nigeria Magazine*, No. 14: 125–9.

DAS GUPTA Ashin, 1979. *Indian Merchants and the Decline of Surat c. 1700–1750*, Wiesbaden.

DAVIS W.A. 1918. 'Present Position and Future Prospects of the Natural Indigo Industry', *Agricultural Journal of India*, Vol. XIII Part 1: 32–46, Part 2: 206–21, Part 3: 441–59, Calcutta.

DEAN Jenny, 1994. *The Craft of Natural Dyeing*, Tunbridge Wells.

DIDEROT Denis, 1765. *Encyclopédie*, Paris, Vols 8 and 22.

DOMONKOS Ottó, 1981. *Blaudruckhandwerk in Ungarn*, Budapest.

DOUMET Joseph, 1980. *Etude sur la couleur pourpre ancienne*, Beirut.

DUSENBERG M. and KUZUKO Takimoto, 1986. *Textiles of Old Japan: Color and Dye*, San Francisco.

Dyes from Nature 1990. Brooklyn Botanic Gardens, Vol. 46/2, New York.

Dye Plants and Dyeing, 1964. Brooklyn Botanic Gardens, Vol. 20/3, New York.

East India Company, 1836. *Reports and documents connected with the proceedings of the East India Company in regard to the culture and manufacture of cotton-wool, raw silk and indigo* (3 parts in 1), London.

EDELSTEIN S.M. and BORGHETTY H.C. (trans. and eds), 1969. *The Plictho of Gioanventura Rosetti*, Cambridge, Massachusetts.

EDMONDS John, 1992. 'The medieval woad trade in England and places of cultivation', *Beiträge zur Waidtagung* 4/5 (ed. MÜLLEROTT H.E), Teil 1: 16–20 and Appendices 1 and 2.

FAWCETT Charles G.H., 1936. *The English Factories in India, 1670–1677*, Vol. 1 (new series), Oxford.

FITZ GIBBON Kate and HALE Andrew, 1997. *Ikat: Silks of Central Asia*, London.

FLOUD P. C., 1960. 'The English Contribution to the Early History of Indigo Printing', *JSDC*, LXXVI, June: 344–9.

FORBES R.J., 1956. *Studies in Ancient Technology*, Vol. 4, Leiden.

FORTUNE Robert, 1846. 'A notice of the *Tien-ching*, or Chinese Indigo', *Journal of the Royal Horticultural Society*, London: 269–71.

FOSTER William (ed.), 1968 (reprint). *Early Travels in India, 1583–1619*, New Delhi.

FOSTER William and FAWCETT Charles, 1906–55. *The English Factories in India (1618–84): A Calendar of Documents in the India Office, British Museum and Public Record Office*, 17 Vols, Oxford.

FRASER-LU Sylvia, 1988. *Handwoven Textiles of South-East Asia*, Oxford.

GAVIN Traude, 1996. *The Women's Warpath: Iban Ritual Fabrics from Borneo*, Los Angeles.

GERBER Fredrick H., 1977. *Indigo and the Antiquity of Dyeing*, Florida.

GHOSH A.K., 1944. 'Rise and Decay of the Indigo Industry in India', Vol. IX, No. 11, May: 487–93 and No. 12, June: 537–42, *Science and Culture*, Calcutta.

GILLET J.B., 1958. *Indigofera (Microcharis) in Tropical Africa*, Kew Bulletin Additional Series I, London.

GILLOW John, 1995. *Traditional Indonesian Textiles*, London.
GILLOW John and BARNARD Nicholas, 1996 (reprint). *Traditional Indian Textiles*, London.
GIOBERT G.A., 1813. *Traité sur le Pastel et l'Extraction de son Indigo*, Paris.
GITTINGER Mattiebelle, 1982. *Master Dyers to the World*, Washington DC.
GITTINGER Mattiebelle (ed.), 1989. *To Speak with Cloth: Studies in Indonesian Textiles*, Los Angeles.
GITTINGER Mattiebelle, 1990. *Splendid Symbols: Textiles and Traditions in Indonesia*, Oxford (first published in Washington DC 1979).
GOODWIN Jill, 1982. *A Dyer's Manual*, London.
GOTO Shoichi. *A Bibliography of Dyeing, Printing and Weaving in Japan* 1420–1926 (produced by Ryosho library).
GRAY Lewis Cecil, 1958 (reprint). *History of Agriculture in the Southern United States to 1860*, 2 Vols, Gloucester, Massachusetts.
GRIERSON Su, 1986. *The Colour Cauldron*, Perth.
HAKLUYT Richard, 1965 (facsimile). *The Principall Navigations Voiages and Discoveries of the English Nation (imprinted at London, 1589)*, 2 Vols, Cambridge.
HALD Margrethe, 1980. *(Olddanske tekstiler) Ancient Danish Textiles from Bogs and Burials*, (trans. J. Olsen), National Museum Archaeological-Historical Series Vol. XXI, Copenhagen.
HALL Allan R., 1992. 'Archaeological records of woad (*Isatis tinctoria L.*) from medieval England and Ireland', *Beiträge zur Waidtagung* 4/5 (ed. MÜLLEROTT H.E.), Teil 1: 21–2 and tables 1 and 2.
HALLER R., 1951. 'The Production of Indigo', 'The History of Indigo Dyeing' and 'The Application of Indigo in Textile Printing', *CR* 85: 3072–87.
HALLEUX Robert, 1981. *Les Alchemistes Grecs*, Vol. 1, Paris.
HAMILTON Roy W. (ed.), 1994. *Gift of the Cotton Maiden: Textiles of Flores and the Solor Islands*, Los Angeles.
HARLEY R.D., 1982 (2nd edn). *Artists' Pigments c. 1600–1835*, London.
HARRIS Jennifer (ed.), 1993. *5000 Years of Textiles*, London.
HELLOT Jean, 1750. *L'Art de la teinture des laines et des étoffes de laine en grand et petit teint*, Paris (English translation in *The Art of Dyeing Wool, Silk and Cotton*, London, 1785).
HERINGA Rens, 1989. 'Dye Process and Life Sequence: the Colouring of Textiles in an East Javanese Village', in GITTINGER (ed.) 1989: 106–30.
HILL David J., 1992. 'Production of natural indigo in the United Kingdom', *Beiträge zur Waidtagung* 4/5 (ed. MÜLLEROTT H.E.), Teil 1: 23–6 and tables 1–7.
HITCHCOCK Michael, 1991. *Indonesian Textiles*, London.
HOLMES Jack, 1967. 'Indigo in Colonial Louisiana and the Floridas', *Louisiana History*, 9/4: 329–49.
HORSFALL R.S. and LAWRIE L.G., 1949 (rev. edn). *The Dyeing of Textile Fibres*, London.
HOSKINS Janet, 1989. 'Why do Ladies Sing the Blues?', in WEINER Annette B. and SCHNEIDER Jane (eds), *Cloth and Human Experience*, Washington DC: 141–73.
HOWARD Albert and HOWARD Gabrielle, 1914. 'The improvement of indigo in Bihar' (Bihar Planters' Association), Calcutta.
HUMMEL J.J., 1890. *The Dyeing of Textile Fabrics*, London.
HURRY Jamieson B., 1930. *The Woad Plant and its Dye*, London (reprint, Clifton, NJ, 1973).
Indigo Publications, 1918–23, reports for the Agricultural Research Institute, Pusa, 1–12 (9 by DAVIS W.A.), Calcutta.
Indigo rein, 1907 (2nd edn), BASF (Badische Anilin und Soda Fabrik), Ludwigshafen am Rhein.
Indigo Research Reports 1906–13, Bihar Planters' Association, Calcutta.
IRWIN John and BRETT Katharine B, 1970. *Origins of Chintz*, London.
IRWIN John and HALL M., 1971. *Indian Painted and Printed Fabrics*, Ahmedabad.
JACKSON Anne, 1997. *Japanese Country Textiles* (V & A collection), London.
JAUBERT G.F., 1900. *La Garance et l'Indigo*, Paris.
JOLY N., 1839. *Etudes sur les plantes indigofères en général, et particulièrement sur le Polygonum tinctorium* (Extrait du Bulletin de la Societé d'Agriculture du Département de l'Hérault, Jan./Feb. 1830), Montpellier.
KAHLENBURG Mary Hunt, 1977. *Textile Traditions of Indonesia*, Los Angeles.
KATZENBERG Dena S., 1973. *Blue Traditions: Indigo Dyed Textiles and Related Cobalt Glazed Ceramics, 17th–19th Centuries*, exh. cat., Museum of Art, Baltimore.
KAYE J.W., 1867. *India, Ancient and Modern*, Vol. 2, London.
KLING Blair B., 1966. *The Blue Mutiny: The Indigo Disturbances in Bengal, 1859–1862*, Philadelphia.
KNECHT Edmund and FOTHERGILL James, 1912. *The Principles and Practice of Textile Printing and Dyeing*, London.
KNECHT Edmund, RAWSON Christopher and LOWENTHAL Richard, 1941 (9th edn). *A Manual of Dyeing*, 2 Vols, London.
KRÜNITZ Johann Georg, 1783. *Oekonomische Encyklopädie oder allgemeines System der Staats=Stadt=Haus=und Landwirtschaft in alphabetischer Ordnung*, Berlin.
LAMB Venice and HOLMES Judy, 1980. *Nigerian Weaving*, Hertingfordbury.
LAMB Venice and Alastair, 1981. *Au Cameroun, Weaving-Tissage*, Hertingfordbury.
LAMB Venice and Alastair, 1984. *Sierra Leone Weaving*, Hertingfordbury.
DE LANESSAN J.-L. (ed.), 1886. *Les Plantes Utiles des Colonies Françaises*, Paris.
LARSEN Jack Lenor, BÜHLER Alfred and SOLYOM Bronwen and Garrett, 1976. *The Dyer's Art: ikat batik plangi*, New York.
LASTEYRIE DU SAILLANT C.P., 1811. *Du pastel, de l'indigotier, et des autres végétaux dont on peux extraire une couleur bleue*, Paris.
LAWRIE L.G. 1967 (10th edn). *A Bibliography of Dyeing and Textile Printing*, London.
LEGGETT William F., 1944. *Ancient and Medieval Dyes*, New York.
LEMMENS R.H.M.J. and WULIJARNI-SOETJIPTO N. (eds), 1991. *Plant Resources of South-East Asia No. 3: Dye and tannin-producing plants* (Pudoc/Prosea), Wageningen.

LEWIS Paul and Elaine, 1984. *Peoples of the Golden Triangle*, London.
LU PU, 1981. *Designs of Chinese Indigo Batik*, New York and Beijing.
MACCARTHY F., 1994. *William Morris: A Life for Our Time*, London.
MACLEOD Murdo L., 1973. *Spanish Central America: A Socioeconomic History*, California.
MACQUER Pierre Joseph, 1763. *Art de la teinture en soie*, Paris.
MÄGDEFRAU Werner, 1973. 'Zum waid- und Tuchhandel Thüringischer Städte im späten Mittelalter', *Jahrbuch für Wirtschaftsgeschichte*, 2: 131–48.
MAIRET Ethel M., 1916. *A Book on Vegetable Dyes*, London.
MARSDEN William, 1811 (reprint 1966). *The History of Sumatra*, London.
MARTIN-LEAKE Hugh, 1975. 'An Historical Memoir of the Indigo Industry of Bihar', *Economic Botany*, Vol. 29, No. 4: 361–71.
MAXWELL Robyn, 1990. *Textiles of Southeast Asia: Tradition, Trade and Transformation*, Melbourne.
MCLAREN K., 1986 (2nd edn). *The Colour Science of Dyes and Pigments*, Bristol and Boston.
MENDE Kazuko and REIKO Morishige, 1991. *Sashiko: Blue and White Quilt Art of Japan*, Tokyo.
MIKI Yokichiro (ed.), 1960. *Awa Ai Fu* , 'The Book of Indigo from Awa Province' (in Japanese), Tokushima.
MILBURN William, 1813. *Oriental Commerce*, 2 Vols, London.
MILLER Dorothy, 1984. *Indigo from Seed to Dye*, Indigo Press, Aptos, California.
MITTAL S. K., 1978. *Peasant Uprisings and Mahatma Gandhi in North Bihar (A Politico-Economic Study of Indigo Industry 1817–1917 with special reference to Champaran)*, Meerut.
MOEYES Marjo, 1993. *Natural Dyeing in Thailand*, Bangkok.
MOHANTY B.C., CHANDRAMOULI K.V. and NAIK H.D., 1987. *Natural Dyeing Processes of India*, Ahmedabad.
MOLLA M.K.U., 1973. 'Ryot Revolt of Pabna in 1860 (an account of the Indigo Uprisings)', *Journal of the Asiatic Society of Bangladesh*, 18 (2): 91–108.
MONNEREAU Elias, 1769. *The Complete Indigo-Maker*, London.
MONTGOMERY Florence M., 1970. *Printed Textiles: English and American Cottons and Linen 1700–1850*, London.
MOOK-ANDREAE Hanne, 1985. 'Het wonder van de indigo', in OEI (ed.) 1985: 13–20.
MOZIÑO José Mariano, 1826. *Tratado del xiquilite y a–il de Guatemala*, Manila.
MÜLLEROTT H.E., 1992. *Quellen zum Waidanbau in Thüringen*, Arnstadt.
MÜLLEROTT H.E. (ed.), 1990, 1992, 1993 and 1994. *Beiträge zur Waidtagung* (International Woad Conference papers), Jahrgang 3 and 4/5, Arnstadt.
NAPIER James A., 1875 (3rd edn). *A Manual of Dyeing and Dyeing Receipts*, London.
NAYLOR G., 1986. *William Morris by Himself*, London.
NEEDHAM Joseph, 1986. 'Biology and Biological Technology – Botany', Vol. 6, Part 1 of *Science and Civilisation in China*, Cambridge.
DE NEGRI Eve, 1966. 'Nigerian textile industry before independence', *Nigeria Magazine*, No. 89, June: 95–101.
NIESSEN Sandra A., 1985. 'Waarom het garen van de godin der Toba Batak zwart was', in OEI (ed.) 1985: 137–44.
O'CONNOR Deryn, 1981. 'Red and Blue', *William Morris and Kelmscott*, exh. cat. (West Surrey College of Art and Design, Farnham), London .
O'CONNOR Deryn and GRANGER-TAYLOR Hero, 1982. *Colour and the Calico Printer*, exh. cat. (West Surrey College of Art and Design), Farnham.
OEI Loan (ed.), 1985. *Indigo – Leven in een kleur*, Amsterdam.
OGIKUBO Kiyoko, 1993. *Kogin and Sashiko Stitch*, Kyoto Shoin's Art Library of Japanese Textiles, Vol. 13, Kyoto.
Papers Relating to the Settlement of Europeans in India 1854. Calcutta.
Papers Relating to the Cultivation of Indigo in the Presidency of Bengal, 1860. Calcutta.
PARKINSON John, 1640. *The Theater of Plantes*, London.
PARRY Linda, 1983. *William Morris Textiles*, London.
PARTRIDGE William, 1823 (reprint 1973, Pasold Research Fund, London). *A Practical Treatise on Dying [sic] of Woollen, Cotton and Skein Silk*, New York.
PEEK Marja F.J., 1995, *TINCL: Database voor Kunsttechnologische Bronnen/ Database for Art Technological Sources* (Central Research Laboratory for Objects of Art and Science), Amsterdam.
PELSAERT Franciso, 1925 (original edn c. 1626). *Remonstrantie*, trans. by MORELAND W.H. and GEYL P. as *Jahangir's India*, Cambridge.
PENLEY Dennis, 1988. *Paños de Gualeceo*, Cuenca.
PERSOZ Jean François, 1846. *Traité théorique et pratique de l'impression des tissus*, 4 Vols, Paris.
PETTIT Florence H., 1974. *America's Indigo Blues: Resist Printed and Dyed Textiles of the Eighteenth Century*, New York.
PICTON John, 1980. 'Women's Weaving: the Manufacture and Use of Textiles Among the Igbirra People of Nigeria', in IDIENS Dale and PONTING K.G. (eds), *Textiles of Africa* , London: 63–88.
PICTON John and MACK John, 1989 (2nd edn). *African Textiles*, London.
Piecework, Sept./Oct. 1994 (issue on indigo-dyed textiles).
LE PILEUR D'ALPIGNY, 1776. *Cultures du Pastel, de la Guade et de la Garance*, Paris.
PIROCH Sigrid, 1988. 'Slovak Folk Art: Indigo Blue Printing', *Ars Textrina* 9, July: 63–124.
PLOWRIGHT Charles B., 1903. 'On the archaeology of woad', *Journal of the British Archaeological Association*, Vol. 9: 95–110.
POLAKOFF, Claire, 1982. *African Textiles and Dyeing Techniques*, London (previously *Into Indigo*, New York, 1980).
POMET Pierre, 1694. *Histoire générale des drogues, traitant des plantes, des animaux et des mineraux etc.*, Paris (English version 2 Vols, London 1712).
PONTING K.G., 1976. 'Indigo and Woad', *Folk Life* 14: 75–88.
PONTING K.G., 1980. *A Dictionary of Dyes and Dyeing*, London.
PUYMAURIN Jean Pierre Casimir de Marcassus, 1810. *Notice sur le pastel (Isatis tinctorum), sa culture et les moyens d'en retirer l'indigo*, Paris.
RAFFLES Thomas Stamford, 1994 (original edn 1817). *History of Java* (intro. by BASTIN John), 2 Vols, Oxford and Kuala Lumpur.
RAMBERT Gaston (ed.), 1949–66. *Histoire du Commerce de Marseille*, 7 Vols, Paris.

RAWSON Christopher, 1899. 'The Cultivation and Manufacture of Indigo in Bengal', *JSCI*, Vol. XVIII: 467–74, and *JSDC*, July: 166–174.

REBORA Giovanni (ed.), 1970. *Un Manuale di Tintoria del Quattrocento*, Milan.

REED Peter, 1992. 'The British chemical industry and the indigo trade', *The British Journal for the History of Science*, Vol. 25: 113–25.

REID James *et al.*, 1995. *Textile Art of Peru*, Lima.

REID W.M., 1887. *The Culture and Manufacture of Indigo (with A Description of a Planter's Life and Resources)*, Calcutta.

REIN J.J., 1889. *The Industries of Japan*, London.

REMBERT David H., 1979. 'The Indigo of Commerce in Colonial North America', *Economic Botany*, 33/2: 128–34.

Reminiscences of Behar 1887/8, written by 'An Old Planter', Calcutta/London.

RHIND William, 1877. *A History of the Vegetable Kingdom*, London.

ROBINSON Stuart, 1969. *A History of Dyed Textiles*, London.

ROGERS George C., 1970. *The History of Georgetown County*, Columbia.

ROSETTI Gioanventura, 1548. *Plictho de larte de tentori. . .* (see EDELSTEIN).

ROSSI Gail, 1990. 'Enduring Dye Traditions of China's Miao and Dong People', *Dyes from Nature*, Brooklyn Botanic Garden Records, Vol. 46, No. 2: 20–23.

ROSSIGNON Julio, 1859. *Manual del cultivo del añil y del nopal*, Paris.

Rouge, Bleu, Blanc. Teintures à Nîmes, exh. cat., Musée de vieux Nîmes, 1989.

ROXBURGH William, 1811. *Account of a new species of Merium [sic], the leaves of which yield indigo. . .*, London.

RUBIO Manuel, 1976. *Historia del añil o xiquilite en Centro América*, 2 Vols, San Salvador (Ministry of Education).

RUFINO Patrice Georges, n.d. *Le Pastel: Or bleu du Pays de Cocagne*, Panayrac.

RUMPHIUS Georgius Everhardus, 1750. *Het Amboinsch Kruidboek*, Vol. 5, Amsterdam.

SAINSBURY Ethel Bruce, 1907–22. *A Calendar of the Court Minutes of the East India Company*, 6 Vols covering 1635–63, Oxford.

SALTZMAN Max, 1992. 'Identifying Dyes in Textiles', *American Scientist*, 474–81.

SALZMAN L.F. 1923. *English Industries of the Middle Ages*, Oxford.

SANDBERG Gösta, 1989. *Indigo Textiles*, London.

SAYER Chloë, 1985. *Mexican Textiles*, London.

SCHEVILL Margot Blum, BERLO Janet Catherine and DWYER Edward B. (eds), 1991. *Textile Traditions of Mesoamerica and the Andes*, Texas

SCHREBER Daniel Gottfried, 1752. *Historische, Physische und Ökonomische Beschreibung des Waidtes. . .*, Halle.

SCHRIRE B.D., 1995. 'Evolution of the tribe Indigofera (Leguminosae-Papilionoideae)', in CRISP M. and DOYLE J.J. (eds) *Advances in Legume Systematics* 7: Phylogeny, Kew (Royal Botanic Gardens): 161–244.

SCHUNK Edward, 1855–7. 'On the Formation of Indigo-blue', *The London, Edinburgh and Dublin Philosophical Magazine and Journal of Science*, Part 1: 73–95.

SCHWARTZ Paul R., 1953. 'Contribution à l'histoire de l'application du bleu d'indigo (bleu anglaise) dans l'indiennage européen', *Bulletin de la Société Industrielle de Mulhouse*, No. II, November: 63–79.

SCHWEPPE H., 1992. *Handbuch der Naturfarbstoffe: Vorkommen, Verwendung, Nachweis*, Landsberg/Lech.

SCHWEPPE H., 1997. 'Indigo and Woad', in WEST FITZHUGH E. (ed.), *Artists' Pigments: A Handbook of Their History and Characteristics*, Vol. 3: 80–107, Washington and Oxford.

SEEFELDER Mathias, 1994 (reprint). *Indigo*, Landsberg.

SEILER-BALDINGER Annemarie, 1994. *A Classification of Textile Techniques*, Bathurst.

SHARRER G.T., 1971. 'Indigo in Carolina, 1671–1796', *South Carolina Historical Magazine*, Vol. 72: 94–103.

SINHA N.K., 1965/70. *The Economic History of Bengal*, Vols 1 and 3, Calcutta.

SINHA N.K., 1976. 'Indigo Industry in Bihar during the 19th Century', *Journal of Indian History*, 54/3: 641–8.

SLOANE Hans, 1725. *Natural History of Jamaica*, Vol. 2, London.

SMITH Robert S., 1959. 'Indigo Production and Trade in Colonial Guatemala', *Hispanic American Historical Review*, Vol. 39/2: 181–211.

SOREK C. and AYALON E. (eds), 1993. *Colours from Nature: Natural Colours in Ancient Times*, Tel Aviv.

SPOONER Roy C., 1943. 'Szchewan Indigo' and 'Experiments on the Improvements in Szchewan Indigo', *Journal of the West China Border Research Society*, 14 (Series B): 102–18.

STANFIELD Nancy. 'Dyeing Methods in Western Nigeria', in BARBOUR and SIMMONDS (eds) 1971.

STEINMANN Alfred, 1947. 'Batiks', *CR* 58: 2090–2120.

STEINMANN Alfred, 1949. 'Das Batiken in China', *Sinologica*, Vol. 2, No. 2: 104–26.

STERLING BENJAMIN Betsy, 1996. *The World of Rozome*, Tokyo.

Sublime Indigo, 1987, exh. cat., Musées de Marseille.

Teinture, expression de la tradition Afrique noire, 1982, exh. cat., Musée de l'Impression sur Étoffes, Mulhouse

TENDULKAR D.G., 1957. *Ghandi in Champaran*, New Delhi.

DU TERTRE, Jean-Baptiste, 1658. *L'Histoire générale des Antillesé*, Paris.

THEISON Heide, 1982. 'Herstellung eines Batak-Tuches', *Archiv für Völkerkunde* 36, Museum für Völkerkunde, Vienna: 59–94.

THOMAS Hugh, 1997. *The Slave Trade: The History of the Atlantic Slave Trade: 1440–1870*, New York and London.

THIRSK Joan, 1997. *Alternative Agriculture: A History from the Black Death to the Present Day*, Oxford.

URE Andrew, 1840. *A Dictionary of Arts, Manufactures, and Mines*, New York and London.

VAN LAER G., 1874. *Recueil des principaux procédés de teintures à mordant à l'usage des teinturiers*, Part 3, Verviers.

VARADARAJAN Lotika, 1983. *Ajrakh and Related Techniques*, Ahmedabad.

VARADARAJAN Lotika. 'Indigo, de Indiase traditie', in OEI (ed.) 1985.

VETTERLI W.A., 1951. 'The History of Indigo', *CR* 85: 3066–71.

VYDRA Joseph, 1954. *L'Imprimé indigo dans l'art populaire slovaque*, Prague.

WADA Yoshiko, KELLOGG RICE Mary and BARTON Jane, 1983. *Shibori : The Inventive Art of Japanese Shaped Resist Dyeing* Tokyo, New York and London.

WAHLMAN Maude and CHUTA Enyinna, 1979. 'Sierra Leone Resist-Dyed Textiles', in CORDWELL Justine M. and SCWARZ Ronald A. (eds), *The Fabrics of Culture*, The Hague: 447–66.

WALRAVERS H. (ed.), 1993. *Ein Blaues Wunder: Blaudruck in Europa und Japan*, Berlin

WARMING Wanda and GAWORSKI Michael, 1981. *The World of Indonesian Textiles*, Tokyo.

WATT George, 1890. *Pamphlet on Indigo*, Calcutta.

WATT George (ed.), 1890. *A Dictionary of the Economic Products of India*, Vol. 4, London.

The *Wealth of India*, Series on Raw Materials (Council of Scientific and Industrial Research), Vol. V (H–K), New Delhi, 1959 (reprint 1991).

WEDELIUS Georg Wolfgang, 1675. *De Sale Volatili Plantarum*, Jena.

WEIR Shelagh, 1989. *Palestinian Costume*, London.

WENGER S. and BEIER H.U., 1957. 'Adire – Yoruba Pattern Dyeing', *Nigeria Magazine*, No. 54: 208–25.

WESTON Christine, 1944. *Indigo* (a novel), London.

WHITFIELD Roderick and FARRER Anne, 1990. *Caves of the Thousand Buddhas*, London.

WILLS Norman T., 1970. *Woad in the Fens*, Spalding.

WILSON Verity, 1996 (reprint). *Chinese Dress*, London.

WOLFF P., 1954. *Commerces et Marchands de Toulouse (vers 1350–vers 1450)*, Paris.

WOUTERS Jan, 1992. 'A new method for the analysis of blue and purple dyes in textiles', *DHA* 10: 17–21.

WULFF Hans E., 1966. *The Traditional Crafts of Persia*, Massachusetts.

YING-HSING Sung, 1966. *T'ien-Kung K'ai-Wu : Chinese Technology in the Seventeenth Century*, Pennsylvania and London.

Manuscripts and unpublished works

BEAULIEU MS of 1731–2, Bibliothèque du Musée d'Histoire Naturelle, Paris (193 (1) J 3874).

BOND Virginia, 1996. 'A Study of Bark Cloth from the Solomon Islands – with particular reference to the use of indigo in the north-western region', MA dissertation, Sainsbury Research Unit, University of East Anglia, Norwich.

BYFIELD Judith, 1993. 'Women, Economy and the State: A Study of the Adire Industry in Abeokuta (Western Nigeria), 1890–1935', Ph.D. thesis, University of Columbia.

CARDON Dominique, 1990. 'Technologie de la draperie médiévale d'après la réglementation technique du nord-ouest Méditerranéen (Languedoc–Roussillon–Catalogne–Valence– Majorque), 13è–15è siècles', Ph.D. thesis, Toulouse University, Vol. 2.

East India Company's Court of Directors' original minutes, Oriental and India Office, British Library, London.

FERNANDEZ MOLINA, Jose Antonio, 1992. 'Colouring the World in Blue – the Indigo Boom and the Central American Market, 1750–1810', Ph.D. thesis, University of Texas.

FLOYD Troy S., 1959. 'Salvadorean Indigo and the Guatemalan Merchants: A Study in Central American Socio-Economic History, 1750–1800, Ph.D. thesis', University of California.

ROQUES George, MS on cotton printing in Ahmedabad in 1678, Ref: F.Fr.14614., Bibliothèque Nationale, Paris.

STOKER Kerry, 1997. 'The Cultivation of Woad (*Isatis tinctoria*) for Production of Natural Indigo: Agronomy, Extraction and Bio-Chemical Aspects', Ph.D. thesis, Bristol University.

TOLAT BALARAM Padmini, 1980. 'Natural Indigo and its Use', diploma project, National Institue of Design, Ahmedabad.

Trattato d'Arte della Lana, Cod. 2580, Biblioteca Riccardiana, Florence, f. 151v–152, f. 145v–146v.

Glossary

(Terms not fully explained in the text)

Alizarin Main red colouring matter of madder root (*Rubia tinctorum* L.) and other red dye plants and their synthetic equivalent.

Alkalis Compounds (such as plant ashes, potash, slaked lime and caustic soda) highly soluble in water, that have the ability to neutralize acids.

Aniline Oily liquid compound, colourless when pure. It was isolated in 1826 by distilling natural indigo with lime and discovered in coal tar in 1834. In 1841 it was found that it could also be obtained by heating caustic potash with indigo, and it then received its name, which is derived from *nila* (Sanskrit for indigo), *an-nil* (Arabic) and *anilera* (Portuguese). In 1856 William Henry Perkin revolutionized the dyeing industry by using aniline to produce the first synthetic dye, 'mauveine'. Obtained from coal tar derivatives, it provides the chemical base of many modern synthetic dyes.

Bast fibres Strong fibrous plant material – e.g. hemp, ramie, flax, nettle and banana – which was commonly used for woven cloth and in many areas predated the introduction of cotton.

Blockprinting Method of printing a pattern on cloth (with a resist or mordant) before dyeing, using a block, usually of carved wood, or with a raised design of metal strips or pins. Sometimes a thickened dye is directly printed.

Cochineal Red insect mordant dye, obtained mainly from the parasite *Dactylopius coccus O.Costa*, which lives on prickly cacti plants.

Copperas Hydrated ferrous sulphate/green vitriol used in the 'copperas' indigo vat.

Fermentation vat Until the eighteenth century indigo dyeing relied on fermentation for the chemical changes needed to transform insoluble indigo into a soluble leuco-form by gradual reduction in an alkaline solution in the dye vat.

Hydrolysis A chemical reaction of a compound caused by decomposition in water; a bond in the compound is split and hydrogen and hydroxyl are added to the fragments.

Indigotin A word sometimes used to describe the insoluble blue dyestuff/pigment indigo. It is produced after the glucoside indican in indigo leaves has hydrolyzed to indoxyl ('indigo white') and glucose; the indoxyl is then converted into indigotin (i.e. indigo) by vigorous beating to add oxygen.

Isatan B An unstable precursor of indigo, contained alongside the more stable but similar indican in the leaves of the woad species.

Kermes Red mordant dye found in insect parasites (*Kermes vermilio Planchon*) of the Mediterranean kermes oak tree. Root of the word 'crimson'.

Kola nut Large nut from the tree *Cola nitida*, formerly widely traded in West Africa. Used as a mild narcotic and gum strengthener, but the 'crushed juice' also used to provide a golden-brown textile dye often combined with indigo resist.

Lac dye Red mordant dye obtained from insects of the *Kerria* family (mainly *Kerria lacca Kerr*), found in Asia. They also create a resinous secretion from which shellac varnish is made.

Logwood Mordant wood dye obtained from the tree *Haematoxylon campechianum* L. Initially confused with indigo, it made an inferior blue but produced especially good blacks, and other colours with various mordants.

Lye Alkaline solution (e.g from plant ashes, caustic soda) used to dissolve indigo, which is insoluble in water with a neutral pH.

Madder Red mordant dye obtained from the roots of *Rubia tinctorum* L. and other species of *Rubia*.

Mordant A chemical (usually a metallic salt) that serves to fix a dye to a fibre by combining with the dye to form an insoluble compound (required for most fast dyes but not for indigo).

Morinda Red mordant dye from the roots of *Morinda citrifolia* L. and *Morinda tinctoria* Roxb., commonly found in Asia.

Orpiment Trisulphide of arsenic, used as a reducing agent in early indigo vats and for 'Pencil Blue' indigo printing in Europe.

Pastel Originally the term used to describe woad dye paste (French *pâte*, Italian *pastello*) but came (in French) to mean the plant itself. Woad paste was mixed with calcium carbonate or gum to use for drawing – hence the name pastel for coloured pastel crayons generally and for the technique of using them.

Pigment Colour which is not fully absorbed as a dye, but which binds to a surface when incorporated in an emulsion. In indigo's case the dried pigment becomes a dye when dissolved in the dye vat with alkalis and a reducing agent.

Prussian Blue Probably the first artificial pigment, a strong blue discovered around 1705 in Germany. From the 1730s widely used for paint and in the next century developed as a dye, the first alternative to indigo.

Reduction The chemical process of removing oxygen from a substance. In the case of vat dyes, including indigo, it is known as the 'vatting' process (see also 'vat dyes', below).

Soga Rich brown dye obtained from the bark of the 'Yellow flame' tree *Peltophorum pterocarpum* D.C. Commonly used in combination with indigo.

Synthetic indigo Based on by-products of the petrochemical industry and made in several ways, two of which are aniline with formaldehyde and sodium/hydrogen cyanide (Prussic acid), and toluene with anthranilic acid.

Vat dyes Fast dyestuffs which are insoluble in water but form compounds soluble in alkalis when reduced (i.e. oxygen removed). In this state dyestuff is deposited on fibres in the dye vat and on contact with the air reverts to insoluble, stable compounds. Indigo and shellfish purple are the only natural vat dyes.

Weld Yellow mordant dye from the European plant *Reseda luteola* L.

Illustration Acknowledgements

Abbreviations

BM (Trustees of the British Museum)
OIOC, BL (Oriental and India Office Collections, British Library)
BASF (Badische Anilin Soda Fabrik company, Ludwigshafen, Germany)
JBP (Jenny Balfour-Paul)

Illustrations are reproduced by courtesy and kind permission of the following:

half-title page: JBP
title page: BM (Eth. 1934 3–7. 238 and 246)
page vii: JBP
page viii: JBP
page 3: The Metropolitan Museum of Art, Gift of John D. Rockefeller, Jun. The Cloisters Collection (37.80.2). Photograph © 1985 The Metropolitan Museum of Art, New York
page 4: JBP
page 6: Widad Kawar Collection, Amman, Jordan. Photo JBP
page 7: Sarabhai Foundation, Calico Museum of Textiles, Ahmedabad
page 8, left: BM (Eth. 1949 As.9.3)
page 8, above right: the late Nancy Stanfield
paes 8–9: Photo collection of Rijksmuseum voor Volkenkunde, Leiden (536–1186). Photo C. Ouwehand
page 10: Photo C. Riesterer, Geschwenda, Germany
page 13: BM (EA 40923)
page 16: Ashmolean Museum, Oxford (1939.590), tempera copy by Nina Davies
page 17: BM (WAA 82–9–18, 2757/83–1–21, 141)
page 18: BM (OA MAS 820)
page 19: Photo Jon Thompson, London
pages 20–21: Otis Norcross Fund, Courtesy of Museum of Fine Arts, Boston (36.625)
page 22: BM (Eth. 1913.3–11.1)
page 24, left: Abegg-Stiftung, Riggisberg, Switzerland (1397)
page 24, right: Coptic Museum, Cairo
page 25: The Whitworth Art Gallery, The University of Manchester (T.8549)
page 26: Rijksmuseum voor Volkenkunde, Leiden (C.71–180)
pages 28–9: Staatliche Museum zu Berlin, Museum für Islamische Kunst (I 5563), photo Jürgen Liepe
page 30: Ashmolean Museum, Oxford (Newberry Collection 1990–250, Cat. 241)
page 31: JBP
page 32: Map by ML Design after J. B. Hurry, *The Woad Plant and its Dye*, London 1930, Fig.11
page 33, above: ADAGP (Societé des Auteurs dans les Arts Graphiques et Plastiques; 91 80 279 V)
page 33, below: Archives départementales du Nord, Musée 342, Cl.J.–Y.Populu (4 Fi 30–381)
page 34: Stadtarchiv Erfurt, photo JBP
page 35: Stadtarchiv Erfurt, photo JBP
page 37: Hampshire Record Office, Winchester (44M69/L30/76)
page 39: Courtesy Hansjürgen Müllerott
page 40: JBP
page 43: The Whitworth Art Gallery, The University of Manchester (T.10119, T.10122)
page 47: OIOC, BL: Court book vol. XXI, page 154
page 49: Courtesy of the Trustees of the V&A, London (IM 160–1929)
page 50: Bibliothèque Centrale du Muséum National d'Histoire Naturelle, Paris (MS 193 (1))
page 51: Bayerische Staatsbibliothek, Munich (Ar.464, fol.138r.)
page 52: JBP
page 53: Musée de l'Impression sur Étoffes, Mulhouse (Lyon 954.60.2)
page 56: Courtesy Hansjürgen Müllerott.
page 58: *Description de l'Egypte* , E. F. Jomard (ed.), Paris 1809–28 (Etat Moderne, vol.II, plate XVI, fig.1)
page 59: Wellcome Institute Library, London: Pierre Pomet, *Histoire générale des drogues*, Paris 1694
pages 60–61: JBP
page 63: Archives of S. Cristóbal de las Casas, Chiapas, Mexico. Courtesy Ana Roquero, photo Antón Laguna
page 65: Michael Bischoff
page 67: Wellcome Institute Library, London: de Beauvais-Raseau, *L'Art de l'Indigotier*, Paris 1770, plate VII
page 71: British Library Board (WD 1017)
page 72–3: Papers Relating to the Settlement of Europeans in India, Calcutta 1854, Appendix I (V 3244)
page 74: British Library Board (NC 688/2)
page 77: Deutsches Textilmuseum, Krefeld, Germany (16044)
page 79: BM (Eth. 1923 4–10.15)
page 80, above: Reproduced by permission of Durham University Library, Wingate Collection (SAD/A27/26)
page 80, below: BM (Eth. 1886.6–28.1)
page 80–81: BM (Eth.1949 Af 46 691)
page 82, above: Courtesy BASF (actual letter in the Deutches Museum, Munich:207)
page 82, below: BASF
page 85: BASF
page 86: Levi Strauss & Co. Archives
page 87: BASF
page 88: JBP
page 90. Map by ML Design
page 91, above: Wellcome Institute Library, London: de Beauvais-Raseau, *L'Art de l'Indigotier*, Paris 1770, plate I
page 91, below: JBP
page 92–3: JBP
page 93, below: Courtesy Hansjürgen Müllerott
page 94–5: JBP
page 94, below: JBP
page 95, right: JBP
page 96–7: the late Nancy Stanfield
page 98: JBP
page 101: Michael Bischoff
page 103, left: JBP
page 103, right: Paolo Costa
page 104: Reproduced by permission of the Trustees of the Wisbech and Fenland Musuem, Wisbech, Cambridgeshire
page 104–5: Reproduced by permission of the Trustees of the Wisbech and Fenland Museum, Wisbech, Cambridgeshire
page 105, below: JBP
page 106, left: JBP
page 106, right: the late Nancy Stanfield
page 107: the late Nancy Stanfield
page 108, above: JBP
page 108, below: JBP
page 109: Coralie Hood
page 111: Journal of the Society of Dyers and Colourists, July 1899, vol. XV, between pages 176 and 177
page 114: JBP
page 116: Kenneth Seddon and Timothy A. Evans, School of Chemistry, The Queen's University of Belfast
page 117: JBP
page 118, above: BM (Eth. 1927.3–10.59)
page 118, below: JBP

page 120, left: the late Nancy Stanfield.
page 120, right: JBP
page 122: JBP
page 123, above: JBP
page 123, below: Antón Laguna and Ana Roquero
page 124: JBP
page 125: JBP
page 126: JBP
page 127: Sandra Niessen
page 130: JBP
page 131: Móra-Ferenc Museum (62.61). Photo Szerencés CBC
page 132: JBP
page 133: JBP
page 134: Maria and Pascal Maréchaux
page 135: Roger Balsom
page 136, left: JBP
page 136, right: Ann Evans
page 137, above left: JBP
page 137, above right: Liszt-Ferenc Museum, Sopron (56.113.1). Photo Szerencés CBC
page 137, below: JBP
page 139: JBP
page 140: Archives Départementales Hérault, Montpellier, France (C 2379). Photo JBP
page 141: David and Margaret Redpath, 'Renaissance Dyeing', Wallis Woollen Mill, Wales. Photo JBP
page 142: JBP
page 143: BM (Eth. 1934.3–7.62)
page 144: JBP
page 145: JBP
page 146: BM (Eth. 1964 Af 2 46)
page 148: JBP
page 149: JBP
page 151: BM (Eth. 1923. 4–10.19)
page 152, above: BM (Eth. 1923.4–10.19)
page 152, below: JBP
page 153: JBP
page 154, left: Exeter University Fine Art Collection
page 154–5: JBP
page 156: JBP
page 157: JBP
page 158, above: Bibliothèque Centrale du Muséum National d'Histoire Naturelle, Paris (MS 193 (1))
page 158, below: Bibliothèque Centrale du Muséum National d'Histoire Naturelle, Paris (MS 193 (1))
page 159: JBP
pages 160–61: Musée de l'Impression sur Étoffes, Mulhouse, France (1624 86–87)
page 161, below: JBP
page 163: Courtesy Finella Balfour–Paul. Photo JBP
page 164, above: Tate Gallery, London (AO 1049)
page 164, below: The Whitworth Art Gallery, The University of Manchester (T. 11418)
page 165: Courtesy Musée de Vieux Nîmes, France. Photo JBP
page 166, above: JBP
page 166, below: Antón Laguna and Ana Roquero
page 167: Ashmolean Museum, Oxford (Shaw Collection X3977)
page 168: BM (Eth. 1980 Am 27.314)
page 169, left: Bobbie Cox
page 169, right: Collection of the late Nancy Stanfield. photo JBP
page 170, above: BM (Eth. 1934. 3–7, 324, 327, 344, 349, 351)
page 170, below: Courtesy Hiroyuki Shindo. Photo T. Hatakeyama
page 172: BM (Eth. 2797)
page 173: Dar al-Tifl Collection, Jerusalem (37.96). photo BM
page 174: BM (JA 1906.12.20.779)
page 175, above left: JBP
page 175, above right: Collection JBP. photo PK, Exeter
page 175, below: JBP
page 176: BM (As 1. 571, 578)
page 178: JBP
page 179: John Gillow
page 180: BM (As 1980 8.1)
page 181: BL (Or 9718 39R)
page 182, left: Fiona Kerlogue, Centre for South-East Asian Studies, University of Hull
page 182–3: Dept. of Anthropology, Smithsonian Institution, Washington DC (361025)
page 184: John Gillow
page 185: BM (Eth. 1949 As 94)
page 186–7: Courtesy of the Trustees of the V&A, London (T. 189. 1948)
page 197, right: John Picton
page 188, above left: John Picton
page 188, below: Venice Lamb
page 188–9: BM (Eth. 1984 Af 2.1)
page 189, below left: Venice Lamb
page 189, below right: Bernard Gardi
page 190: BM (Eth. 1983 Am 35.11)
page 192–3: BM (JA 1945.11–1.047, 1 and 2)
page 194: Balfour-Paul Collection. Photo PK, Exeter
page 195: Susan Conway
page 196: National Maritime Museum, Greenwich, London (BCH 2619)
page 198, above: JBP
page 198, below: Pitt Rivers Museum, University of Oxford
page 199: JBP
page 200: Roger Balsom
page 201: the late Nancy Stanfield
page 202: Chloë Sayer
page 204: Photo Martine Nougarède
page 205: Levi Strauss & Co. Archives
page 206: BL (Or 6729)
page 211, left: Judith Wetherall
page 211, right: Salisbury Cathedral. Photo Eddie Sinclair (S42)
page 213: BM (OA 1983.10–8.1)
page 214: Sotheby's, London
page 216: Collection of Parviz and Manijeh Tanavoli
pages 218–19: Royal Botanic Gardens, Kew, London
page 219, above: Topkapı Palace Museum, Istanbul (A.2147, fol.306a)
page 220: JBP
page 221: JBP
page 222: Maria and Pascal Maréchaux
page 223: Punch Picture Library (cartoon by Arnold Roth)
page 224: Maria and Pascal Maréchaux
page 225: Maria and Pascal Maréchaux
pages 226–7: JBP
page 228: Robert Harding
page 210: Collection JBP. Photo PK, Exeter
page 234: Courtesy Chris Cooksey, University of London

Index

(page numbers in italics refer to illustrations)